Introducing General Relativity

Mark Hindmarsh

University of Sussex
Brighton, UK
    and
University of Helsinki
Helsinki, Finland

Andrew Liddle

University of Edinburgh
Edinburgh, UK
    and
Perimeter Institute for Theoretical Physics
Waterloo, Canada
    and
Universidade de Lisboa
Lisboa, Portugal

This edition first published 2022
© 2022 John Wiley and Sons Ltd

The right of Mark Hindmarsh and Andrew Liddle to be identified as the authors of this work has been asserted in accordance with law.

Registered Offices
John Wiley & Sons, Inc., 111 River Street, Hoboken, NJ 07030, USA
John Wiley & Sons Ltd, The Atrium, Southern Gate, Chichester, West Sussex, PO19 8SQ, UK

Editorial Office
The Atrium, Southern Gate, Chichester, West Sussex, PO19 8SQ, UK

For details of our global editorial offices, customer services, and more information about Wiley products visit us at www.wiley.com.

Wiley also publishes its books in a variety of electronic formats and by print-on-demand. Some content that appears in standard print versions of this book may not be available in other formats.

Library of Congress Cataloging-in-Publication Data applied for

ISBN: 9781118600719

Cover Design: Wiley
Cover Images: © ESA/Hubble & NASA, S. Jha;
Acknowledgment: L. Shatz

Set in 10/12pt NimbusRomNo9L by Straive, Chennai, India

Contents

Preface

Find a physicist. Sit them down, treat them to a coffee, and ask for their opinion as to the most beautiful theory that the subject has to offer. The chances are that their answer will be General Relativity.

Einstein's General Theory of Relativity, to give it its full and proper name, is over a hundred years old. Yet it offers a defiantly modern viewpoint, defining principles of how the Universe ought to work and establishing a mathematical framework upon them. It gives a radical, even shocking, reconception of a fundamental force, gravity. And it has maintained an exquisite agreement with observational data, making predictions of such subtlety that one of its key implications, the existence of gravitational waves, took over a hundred years to be directly verified.

This book is based on a lecture course at the University of Sussex, given by each of us at various times. The course is taken by final-year physics undergraduates, and has no special prerequisites, so we have attempted to limit the coverage of topics and to be as explicit as possible. For many undergraduates, a lecture course on General Relativity is the pinnacle of their theoretical education, and their main exposure to the modern methodology of physics as based on principles and symmetries.

Our aim is to make that experience as enjoyable as possible, while accepting that the pleasure comes not just from the astonishing physical implications, such as black holes, singularities, and gravitational waves, but from the elegance of the underlying mathematical structure. By the standards of textbooks on the topic, we have sought to create something that is genuinely introductory, yet which provides the mathematical tools to see the theory work in a quantitative way. We hope that something of the elegance of the theory emerges from the technical difficulty, along with an understanding of the physical predictions that have been so beautifully confirmed by decades of experiment and observation. Enjoy the challenge!

Mark Hindmarsh and Andrew Liddle
Helsinki and Lisbon, October 2021

Some Fundamental Constants and Astronomical Values

Newton's gravitational constant	G	$6.672 \times 10^{-11}\,\mathrm{m^3\,kg^{-1}\,s^{-2}}$
Speed of light	c	$2.998 \times 10^{8}\,\mathrm{m\,s^{-1}}$
Reduced Planck constant	$\hbar = h/2\pi$	$1.055 \times 10^{-34}\,\mathrm{m^2\,kg\,s^{-1}}$
Boltzmann constant	k_B	$1.381 \times 10^{-23}\,\mathrm{J\,K^{-1}}$
Solar mass	$M_\odot$	$1\,M_\odot = 1.989 \times 10^{30}\,\mathrm{kg}$
Solar radius	$R_\odot$	$1\,R_\odot = 6.957 \times 10^{8}\,\mathrm{m}$
Parsec	pc	$1\,\mathrm{pc} = 3.086 \times 10^{16}\,\mathrm{m}$
Electron volt	eV	$1\,\mathrm{eV} = 1.602 \times 10^{-19}\,\mathrm{J}$

Commonly-Used Symbols

About the Companion Website

This book is accompanied by a companion website:
www.wiley.com/go/hindmarsh/introducingGR

The website includes: Instructor's Manual

Chapter 1

Introducing General Relativity

It is now more than a hundred years since Albert Einstein presented the final form of the General Theory of Relativity to the Prussian Academy of Sciences, in November 1915. Since then, it has migrated from an extraordinary achievement at the frontiers of physics, reputedly understood by only a very few, to a standard advanced undergraduate course. General Relativity (as it is usually called, commonly shortened to simply GR) is essential for the understanding of the Universe as a whole, wherever gravity is strong, and also whenever precise time measurements are made. The Global Positioning System (GPS), now built in to billions of devices around the world, would not work without the General Relativistic prediction that clocks run more slowly on Earth than in the satellites defining the GPS reference frame.

Part of the fascination of General Relativity lies in the personality of Einstein, and the way he is often presented as a lone genius working for years in isolation, finally to reemerge with the fully formed and beautiful theory we know today. In reality, he was in constant communication with other scientists, and others were working towards a relativistic theory of gravitation. The first such theory was actually written down by Gunnar Nordström in 1913, who attempted a direct relativistic generalisation of the Newtonian gravitational potential. Einstein was the first to understand that the appropriate dynamical quantity is the space–time metric itself, but the geometric aspect of General Relativity was probably first appreciated by the mathematicians Marcel Grossmann and David Hilbert. Einstein worked with his friend Grossmann, and had crucial correspondence with Hilbert before coming up with the final and correct formulation. Einstein himself took several wrong turnings doing the years between 1907 and 1915 when he was working most intensively on the theory. The lone genius is a myth, but it is fair to consider that General Relativity is Einstein's own, and crowning, achievement.

The technical complexity of General Relativity comes from several sources.

Introducing General Relativity, First Edition. Mark Hindmarsh and Andrew Liddle.
© 2022 John Wiley & Sons Ltd. Published 2022 by John Wiley & Sons Ltd.
Companion website: www.wiley.com/go/hindmarsh/introducingGR

The fundamental objects of relativity are tensors, because the relativity principles (both special and general) are statements about the properties of physical laws under transformations between coordinates. Thus any General Relativity course must start with tensor calculus. General Relativity is a geometrical theory, treating space–time as a manifold, describing its dynamics in terms of geometrical quantities. Thus in approaching General Relativity the basic geometrical concepts developed by Bernhard Riemann and others from the mid 1800s — of connection, geodesic, parallel transport, and curvature — must be introduced. Tensor calculus and Riemannian geometry are not part of the standard mathematical equipment of a physics undergraduate. This was also true in Einstein's undergraduate career, although there were courses on offer. It has been speculated that had he gone to an advanced geometry course, he would have later saved himself several years' work.

A final difficulty is the one of translating the mathematical concepts into physical observables. General Relativity rethinks the fundamentals of space and time, which take part in physical processes rather than being a framework on which things happen. So deciding what is observable, rather than simply an artefact of a particular choice of coordinates, is difficult. Indeed, Einstein changed his mind a couple of times as to whether gravitational waves were real or not, and it took about fifty years for a unanimous view to emerge.

Gravitational waves, an early prediction of General Relativity, are amongst the hottest topics in physics following their direct detection by the LIGO/Virgo collaboration, announced in 2016. Their real significance is not so much as a triumphant vindication of Einstein's theory; there was no serious doubt that gravitational waves existed following the careful measurements of the orbital decay of a binary pulsar system discovered by Russell Hulse and Joseph Taylor in the 1970s. Rather, the detection signals the beginning of a new branch of astronomy, which has the prospect of detecting violent astronomical events right back to the very earliest stages of the Big Bang. New detectors, similar to LIGO and Virgo, are being built in Japan and India, and a space-based gravitational wave detector called LISA is planned for the early 2030s. General Relativity will continue to be at the forefront of scientific research in the 21st century, as it was throughout the 20th.

Chapter 2

A Special Relativity Reminder

*The Special Theory unites space and time · shorter lengths and
longer times · seeing it with diagrams*

Before launching into our account of General Relativity, we give a brief reminder of the main characteristics of its predecessor theory, the Special Theory of Relativity. This was introduced by Einstein in 1905, and is usually referred to by the shorthand Special Relativity. These theories have a rather different status to traditional physics topics, such as electromagnetism or atomic physics, which seek to understand phenomena of a particular type or within a certain domain. Instead, the relativity theories set down principles which apply to *all* physical laws and restrict the ways in which they can be put together. Whether those principles are actually true is something that needs to be tested against experiment and observation, but the assumption that they do hold has far-reaching implications for how physical laws can be constructed. In particular, the role of symmetries of Nature is highlighted, which is a defining feature of how modern physics is constructed; as such the relativity theories often give students the first glimpse of how contemporary theoretical physics is done.

Both the theories focus on how physical phenomena are viewed in different coordinate systems, with the underlying principle that the outcome of physical processes should not depend on the choice of coordinates that we use to describe them. Special Relativity restricts us to so-called **inertial frames**, where the term **frame** means a set of coordinates to be used for describing physical laws. As we will see, this restricts us to coordinate transformations which are linear in the coordinates, corresponding to coordinate systems moving relative to one another with constant velocity, and/or rotated with respect to one another. This turns out to be a suitable framework for considering all known physical laws *except for those corresponding to gravity*.

Einstein's remarkable insight, leading to the General Theory of Relativity, was

Introducing General Relativity, First Edition. Mark Hindmarsh and Andrew Liddle.
© 2022 John Wiley & Sons Ltd. Published 2022 by John Wiley & Sons Ltd.
Companion website: www.wiley.com/go/hindmarsh/introducingGR

that allowing arbitrary non-linear coordinate transformations would allow gravity to be incorporated. Indeed, if we want to allow non-linear transformations, we *have* to include gravity. Understanding the motivations for, and implications of, this extraordinary statement is the purpose of this book. But for now, we place the focus on Special Relativity, emphasising those features that will later generalise.

2.1   The need for Special Relativity

In Newtonian dynamics, the equations are invariant under the Galilean transformation which takes us from one set of coordinates (t, x, y, z) to another (t', x', y', z') according to the rule

$$t' = t; \quad x' = x - vt; \quad y' = y; \quad z' = z, \tag{2.1}$$

where v is the relative speed between the two coordinate systems, which have been aligned so that the velocity is entirely along the x direction. [NB primes are not derivatives!] Each coordinate frame is idealised as extending throughout space and time, providing the scaffolding that lets us locate physical processes in space and time. We introduce an **event** as something which happens at a specific location in space and at a specific time, such as the collision of two particles.

Typically any observer will want to choose a coordinate system to describe events, and will be located somewhere within the coordinate system. Commonly, though not always, observers will decide to choose coordinate frames that move along with them as a natural way to describe the phenomena as they see them, and so it can be useful to sometimes think of a coordinate system as being associated to a particular observer who carries the coordinate system along with them. For instance, we might consider two different observers moving at a constant velocity with respect to one another, and ask how they would describe the same physical process from their differing points of view.

When we refer to **invariance** of a physical quantity, we mean that a physical quantity expressed in the new coordinates is identical to the same quantity expressed in the old ones. That means that observers in relative motion agree on its value.

In particular, acceleration is invariant in Newtonian dynamics; it depends on second time derivatives of the coordinates of, for example, a moving particle, and the second time derivatives of x and of x' are equal. An everyday example is that an object dropped in a train moving at constant velocity appears, to an observer in the carriage, to follow exactly the same trajectory as it would were the train stationary.

The Galilean transformation is characterised by a single universal time coordinate that all observers agree upon. Combining relative velocities in each of the coordinate directions means that generally $x' \neq x$, $y' \neq y$, and $z' \neq z$, but t' always remains equal to t. The idea of a universal time sits in good agreement with our everyday experience. However, our own direct perceptions of physical laws probe only a very restrictive set of circumstances. For example, we are unaware of quantum mechanics in our day-to-day life, because quantum laws such as Heisenberg's Uncertainty Principle are significant only on scales far smaller than we can personally witness. Hence, we cannot immediately conclude that invariance under the Galilean transformation should apply to all physical laws.

Indeed, it was already known in Einstein's time that Maxwell's equations, describing electromagnetic phenomena including the propagation of light waves, are not consistent with Galilean invariance. For example, they state that the speed of light is independent of the motion of a source, whereas the Galilean transformation would predict that light would emerge more rapidly from a torch if its holder were running towards you. In a famous thought experiment (i.e. an experiment carried out only in the mind, not in the laboratory), Einstein tried to envisage what would happen if one tried to catch up with a light wave by matching its velocity, knowing that Maxwell's equations would not permit a stationary wave.

One possible resolution of this would be if there were special frame of reference in which Maxwell's equations were valid, a frame that came to be known as the aether. However, since the Earth revolves around the Sun, it cannot always be stationary with respect to this aether. In the late 1880s, Albert Michelson and Edward Morley sought to detect the motion of the Earth relative to this aether, using an interferometer experiment. It should have had the sensitivity to easily see the effect, given the known properties of the Earth's orbit, yet no signal was found, putting the existence of the aether in doubt.

From the viewpoint of wanting a unified view of physical laws, it makes little sense that different types of physical laws should respect different invariance properties. After all, electromagnetic phenomena lead to dynamical motions. This incompatibility posed a stark problem for physics.

Einstein's 1905 paper resolved this seeming paradox decisively in favour of electromagnetism. Based on his thought experiments, he demanded that physical laws satisfied two postulates:

1. **The laws of physics are the same in all inertial frames.**

2. **The speed of light, denoted c, is the same in all inertial frames and independent of the motion of the source.**

As remarked above, inertial frames are those which move with a constant velocity with respect to one another. The requirement that the laws of physics be the same in each is inherited from the Galilean transformation, which also requires it. Another way of expressing this first postulate is to say that there is no possible experiment an observer can carry out to measure their absolute velocity.

But the second postulate then requires that the coordinate transformation between frames must mix space and time, as we are about to see. It is inconsistent with the notion of a universal time coordinate, and requires that invariance under the Galilean transformation be abandoned. If Nature's laws are to be invariant under coordinate transformations, the invariance must be of another type.

2.2   The Lorentz transformation

Hendrik Lorentz, in 1904, had already discovered a transformation that left Maxwell's equations invariant, and it now bears his name. We will derive it under the assumption that the transformation is linear, like a Galilean transformation, and reduces to a Galilean transformation in the limit of relative velocities much less than that of light.

Consider a frame, which we call S', moving relative to the original frame S with velocity v along the x-axis, so that we can assume $y' = y$ and $z' = z$.[1] This is shown in Figure 2.1. Linearity lets us write

$$t' = At + Bx; \quad x' = Ct + Dx, \tag{2.2}$$

where A, B, C, and D are constants. Now, the origin of S' is moving relative to S at velocity v, so $x' = 0$ corresponds to $x = vt$, implying $C = -vD$. So the second of the above equations becomes

$$x' = D(x - vt). \tag{2.3}$$

By symmetry, the same equation must hold for transforming back from S' to S, exchanging $v \rightarrow -v$, so

$$x = D(x' + vt'). \tag{2.4}$$

Now, we use the assumption of a constant speed of light. At the instant when the two coordinate systems agree, send out a pulse of light along the x-axis. Then $x = ct$ in the original frame, and $x' = ct'$ in the new one. Substitute this in to equation (2.3) to get

$$x' = D(1 - v/c)x, \tag{2.5}$$

---

[1] We will refer to such a transformation as a **Lorentz boost**. The general linear transformation satisfying Einstein's postulates, referred to as a Lorentz transformation, also permits rotations of the spatial coordinate frame.

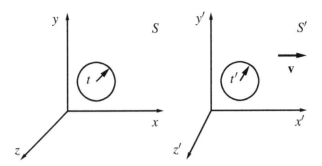

Figure 2.1 A frame S' moving relative to another frame S, with velocity $\mathbf{v}$ along the x-axis. Demanding that the transformation relating the coordinates (t',x',y',z') to (t,x,y,z) is linear and preserves the speed of light c, uniquely fixes it to be the Lorentz boost, equation (2.7).

and into equation (2.4) to get

$$x = D(1 + v/c)x'. \tag{2.6}$$

Consistency of these two requires $D^2 = 1/(1 - v^2/c^2)$. This rules out the Galilean transformation, which would have required D to be equal to one.

We write this as $x' = \gamma(x - vt)$, where the famous γ-factor is $\gamma = 1/\sqrt{1 - v^2/c^2}$. This measures the strength of relativistic effects. Again by symmetry, $x = \gamma(x' + vt')$. Combining these gives $vt' = \gamma^{-1}x - x' = \gamma^{-1}x - \gamma(x - vt)$ which we can re-arrange to get $t' = \gamma(t - vx/c^2)$. We can finally write the Lorentz boost (in the x-direction) as

$$
\begin{aligned}
ct' &= \gamma\left(ct - \frac{v}{c}x\right); \\
x' &= \gamma(x - vt); \\
y' &= y; \quad z' = z.
\end{aligned}
\tag{2.7}
$$

Notice the strong symmetry if we think of ct as a coordinate. The Lorentz boost is the unique linear transformation relating time and space coordinates satisfying the postulates of Special Relativity. It also reduces to the Galilean transformation in the limit $v \ll c$.

If we define the distance of a point in space and time from the origin as

$$s^2 = -c^2t^2 + x^2 + y^2 + z^2, \tag{2.8}$$

this quantity, which is known as the space–time **interval**, is invariant under the

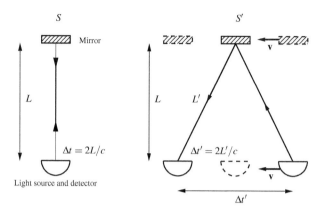

Figure 2.2   Light pulses are sent out from the origin in the y direction in frame S, bouncing off a mirror at a distance L, and returning to the origin at time $t = 2L/c$ later. As soon as they are received another pulse is sent. In frame S', the path is longer, and the arrival times back at the detector correspondingly later. If we think of the arrival times of the pulses as the ticking of a clock, the conclusion is that the moving clock runs slower, by a factor $\gamma = 1/\sqrt{1 - v^2/c^2}$ (see equation (2.9)).

Lorentz boost (see Problem 2.1). Hence, all inertial observers agree on the interval between two events, even though they will not generally agree on how much of that interval is in the space direction and how much in the time direction. Invariant quantities such as this will play an important role throughout our book. Indeed, a more modern and fundamental point of view would be to start with the principle that the interval is to be invariant, and to show that the most general linear transformation consistent with that property is the Lorentz transformation.

2.3   Time dilation

A consequence of the Lorentz transformation is that moving clocks run more slowly. Consider a train of pulses of light sent out from the origin of a frame S, reflected from a mirror a distance L away, and returning a time Δt later. As soon as a pulse returns another is sent.

Consider now how this looks in frame S', moving parallel to the mirror with speed v (see Figure 2.2). In the latter frame, the path taken by the light is longer, and due to the constancy of the speed of light in all frames, the time taken to return $\Delta t'$ is longer.

In Frame S, the pulse return time is $\Delta t = 2L/c$. In Frame S', it is $\Delta t' = 2L'/c$, where $L'^2 = L^2 + (v\Delta t'/2)^2$. Combining these gives

$$\Delta t' = \frac{\Delta t}{\sqrt{1 - v^2/c^2}}. \qquad (2.9)$$

Consider the train of light pulses as the ticking of a clock in the S frame. As $\Delta t' > \Delta t$, the S' observer (who sees a moving clock) notices that its ticks are slower. The situation is symmetric, so whenever we look at a moving clock, we will see it running slower than our own.

A classic example of time dilation comes from muons created in the upper atmosphere through impact of highly energetic cosmic rays. Muons are unstable particles whose lifetime, as measured in the laboratory, is only 2.2×10^{-6} seconds, which suggests that they could travel at most $ct \simeq 660$ m before decaying. But in practice, muons created high in the atmosphere are detected at the Earth's surface. The reason is that they are moving close to the speed of light and hence, from our perspective, their evolution is heavily time-dilated with the apparent lifetime much greater than the rest-frame lifetime.

2.4 Lorentz–Fitzgerald contraction

A complementary phenomenon to time dilation is that moving objects appear shorter, known as the Lorentz–Fitzgerald contraction. To derive it, take a rod that is stationary in frame S with ends at $x = 0$ and $x = L$, and choose a moving frame S' such that the spatial origins coincide at $t = t' = 0$. In this frame, the rod appears to move with speed v in the $-x'$ direction. What is the length of the rod in the moving frame?

To answer the question, we need to be a bit more precise about how we determine lengths. We can define the length of the object to be the difference in the x coordinate between the ends, when measured at the same time. In the S frame, where the problem was set up, this is trivially L, at any time t.

However, in the S' frame, the calculation is not so trivial, as we need to establish the positions of the ends of the rod at the same value of t'. It is convenient to take this time to be $t' = 0$, as we set up the coordinate systems so that their origins coincide at this time. The Lorentz transformation (2.7) tells us that measurements taken at time $t' = 0$ must happen at times $t = xv/c^2$ in the frame S. Hence, the measurement the position of the end of the rod ($x = L$) at $t' = 0$ in the frame S' occurs at time $t = Lv/c^2$ in frame S (see Figure 2.3).

The Lorentz transformation (2.7) of the coordinates corresponding to the mea-

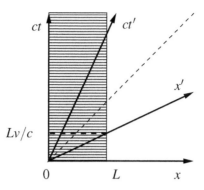

Figure 2.3   A rod of length L is stationary in a frame with coordinates (ct,x), with its left-hand end at the origin. A frame is moving at speed v in the $+x$ direction, with coordinates (ct',x') coinciding with the original frame at $t'=0$ and $x'=0$. A measurement of the position of the end of the rod at time $t'=0$ happens at time $t=Lv/c^2$ in the original frame. Knowing that the end of the rod is at $x=L$, the position of the end of the rod in the moving frame at time $t'=0$, and hence its length L', can be calculated from the Lorentz transform (2.7). The conclusion is that $L'=L/\gamma$, i.e. a moving rod is shorter.

surement $(ct,x)=(Lv/c,L)$ tells us that it happened at S' coordinate

$$x' = \gamma L \left(1 - \frac{v^2}{c^2} \right).$$
(2.10)

Hence, the length of the rod in the moving frame is

$$L' = L\sqrt{1-v^2/c^2}.$$
(2.11)

The rod is therefore shorter, when measured in the moving S' frame.

Returning to the example of the atmospheric muons of the previous section, in the muon rest-frame the lifetime to decay is indeed 2.2×10^{-6} seconds. Nevertheless, muons can still reach the ground from the upper atmosphere, which from their point of view is because the atmosphere is approaching them at relativistic speeds and is Lorentz–Fitzgerald contracted. Ultimately, the physical outcome, muons reaching the ground and being detected there, has to be the same from either our point of view or the muons. Hence, we see that time dilation and Lorentz–Fitzgerald contraction are really two different sides of the same coin; it couldn't make sense to have one without the other.

2.5 Addition of velocities

We now ask how velocities change between moving frames. Consider an object moving with velocity **u** along the x axis in frame S. What is the velocity **u**$'$ measured in frame S', which is moving with velocity **v** along the x axis?

In Newtonian dynamics, where the Galilean transformation applies, velocities simply add. But we can already expect that this will not hold in Special Relativity, because of the postulate that the speed of light is constant in all frames regardless of their relative velocity.

In the S' frame, the component of velocity along the x-axis is given by

$$u' = \frac{\Delta x'}{\Delta t'}, \tag{2.12}$$

where the distance $\Delta x'$ is travelled in a time $\Delta t'$. Using the Lorentz boost given by equation (2.7), we can write (the γ factors immediately cancel out)

$$u' = \frac{\Delta x - v \Delta t}{\Delta t - (v/c^2)\Delta x} = \frac{\Delta x/\Delta t - v}{1 - (v/c^2)\Delta x/\Delta t}. \tag{2.13}$$

Since the x component of velocity measured in frame S is $u \equiv \Delta x/\Delta t$, this becomes

$$u' = \frac{u - v}{1 - (v/c)(u/c)}. \tag{2.14}$$

Going between frames in the opposite direction simply swaps v for $-v$ in the Lorentz transformation, so we can also write the equivalent equation

$$u = \frac{u' + v}{1 + (v/c)(u'/c)}. \tag{2.15}$$

In the non-relativistic limit $v \ll c$ and $u \ll c$, we recover the Galilean law $u' \simeq u - v$ that applies in our everyday experience. Taking $u = c$, we find that $u' = c$ regardless of the value of v, confirming successful implementation of the frame-independence of the speed of light. With a little more work, we can show that as long as both v and u are less than c, then so is u'. There is no way to generate a speed greater than that of light from speeds which are less, so the speed of light forms a barrier that cannot be crossed.

Some further useful properties of the Special Relativistic velocity composition law are explored in Problem 2.4.

2.6 Simultaneity, colocality, and causality

We begin with a few definitions. We recall that an event refers to something which takes place at a specific location in space at a particular time. The set of all possible events is the **space–time**. Events are said to be **simultaneous** if they happen at the same time, and to be **colocal** if they happen at the same location. Note that different observers do not agree on which events are simultaneous or colocal. **Coincident** events happen at the same time and place, and everyone agrees on that.

Consider two events. Unless they are simultaneous, there is always a Galilean transformation which makes them happen at the same location, while in contrast there is no such transformation that can make two events appear simultaneous unless they are so in all frames. In Special Relativity, the situation is more complicated.

Consider the following question. Suppose an observer S sees event A, at location x_A, happening before event B at x_B (i.e. $t_B - t_A > 0$). We can assume both x_A and x_B are on the x-axis with $x_B > x_A$. What condition needs to be satisfied so that *all* observers see event A happening first?

We use the Lorentz transformation

$$ct'_B - ct'_A = \gamma \left[ct_B - ct_A - \frac{v}{c}(x_B - x_A) \right]. \tag{2.16}$$

This is positive provided $ct_B - ct_A > (v/c)(x_B - x_A)$. So, if $(x_B - x_A)$ is not too large, the order of events is observer-independent,

In the new frame, the distance between the events is

$$x'_B - x'_A = \gamma [x_B - x_A - v(t_B - t_A)]. \tag{2.17}$$

We see that if event A is able to cause event B, then we can always find a frame in which $x'_B - x'_A = 0$, that is, there is a frame in which the events are colocal.

Conversely, if $ct_B - ct_A < (v/c)(x_B - x_A)$, we can always find a frame in which $ct'_B - ct'_A = 0$. In this frame, the events are simultaneous.

A crucial notion is **causality**. Event A can only influence event B if it can send a signal, which cannot travel faster than light. Event A can therefore cause event B only if

$$\frac{x_B - x_A}{ct_B - ct_A} \leq 1. \tag{2.18}$$

This is exactly the condition as to whether there is a well-defined order, as no observer can have $v > c$. Relativity therefore respects causality; if it is possible for event A to physically influence event B, then all observers will agree that event

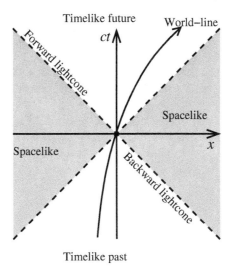

Figure 2.4 Diagram of Minkowski space–time, showing how the light cone divides it into regions separated from the origin by timelike and spacelike (shaded) intervals. Also shown is the world–line of a particle which is steadily accelerating in the $+x$ direction.

A happened first. The general statement is that event A can influence event B if the space–time interval satisfies $s^2 < 0$. Note that s can be imaginary.

2.7 Space–time diagrams

We recall that an event occurs a given point in space–time, and an observer specifies it by the values of four coordinates. There is a useful graphical way of representing events and the causal relationship between them, called a space–time diagram, illustrated by Figure 2.4 for a simplified situation with one space coordinate. The convention is to scale the axes so that light travels at a 45° angle.

The collection of all possible paths taken by light rays emanating from and arriving at a space–time point $\mathscr{O}$ is known as the **light cone**. It consists of all points satisfying $s = 0$, where s is the interval defined above. Note that each point in space–time has its own light cone.

The light cone divides space–time into three regions, **timelike future, timelike past**, and **spacelike**. Only events in the timelike past of $\mathscr{O}$ can influence it, and it can only affect points in its timelike future. Space–time points on the light cone have **lightlike** intervals.

Timelike events are separated by an interval $s^2 < 0$, spacelike ones have $s^2 > 0$, and lightlike ones have $s^2 = 0$. The interval is invariant under a Lorentz transformation: therefore, all observers agree on whether two events are timelike, spacelike, or lightlike separated.

From the earlier discussion in this section, we can see that timelike separated events are colocal in some frame, while spacelike separated events can be made simultaneous.

A **world line** is a path through space–time taken by any particle or observer. A physical particle never exceeds the speed of light, and so its world line can never be inclined more than $45°$ to the time axis. Its world line must always remain within its own forward light cone.

Problems

2.1* The space–time interval between a point at the origin and another at (t, x, y, z) is defined as

$$s^2 = -c^2 t^2 + x^2 + y^2 + z^2.$$

Show that it is invariant under a Lorentz transformation along the x-axis.

[More challenging is to *assume* the invariance of the interval, and show that the Lorentz transformation is the most general linear transformation consistent with that. Try it, for instance restricting to transformations mixing just t and x.]

2.2* A muon, with half-life $t_{1/2} = 2.2 \times 10^{-6}$ seconds, is created in the atmosphere 30 km above the Earth. Assuming it has no further interactions and is travelling directly towards the Earth, what is the minimum velocity (in units of the speed of light) it must have so that it has at least a 50% chance of reaching the Earth's surface before decaying?

2.3 An astronaut has bought a new spaceship capable of travelling near the speed of light. However, they failed to notice that, at 20 m long, it is 4 m longer than their garage. They decide to take advantage of Lorentz–Fitzgerald contraction to fit it in, and ask a friend to close the door behind them.

a) How fast must they travel for the ship to fit into the garage (as seen by their friend)?

b) What is the length of the garage as seen by the astronaut?

c) The friend slams the door as soon as the ship is inside. Afterwards, reviewing their carefully placed array of cameras with synchronised clocks, the friend finds that at the same instant the front crashed through the far wall. What was the sequence of events according to the astronaut?

2.4* The velocity composition law in Special Relativity, equation (2.15), is

$$u = \frac{u' + v}{1 + (v/c)(u'/c)} .$$

Define a new velocity parameter w by $u/c = \tanh w$ (and correspondingly $u'/c = \tanh w'$ and $v/c = \tanh W$); this parameter is known as the **rapidity**. Its range goes from $\pm\infty$, while the velocities go from $-c$ to $+c$.

a) Demonstrate that rapidities obey the addition law $w = w' + W$. In Special Relativity, it is the rapidities that add linearly, not the velocities.

b) Show that in terms of the rapidity, the Lorentz boost can be written as

$$
\begin{aligned}
ct' &= ct \cosh W - x \sinh W ; \\
x' &= -ct \sinh W + x \cosh W ; \\
y' &= y ; \quad z' = z .
\end{aligned}
$$

This shows that the Lorentz boost is equivalent to a hyperbolic rotation (i.e. a rotation through an imaginary angle) in space–time.

c) Show that invariance of the interval under Lorentz boosts follows from the identity $\cosh^2 W = 1 + \sinh^2 W$.

2.5 Draw a space–time diagram showing:

a) The world line of a stationary object.

b) The world line of a massive object moving at a constant velocity.

c) The world line of an object moving with simple harmonic motion on the x-axis.

d) The time and space axes (t', x') of a frame moving with speed u in the $+x$ direction.

2.6 We found that a moving object of rest length L has length L/γ, where $\gamma = \left(1 - v^2/c^2\right)^{-1/2}$. To figure out what the object actually looks like, we need to take into account the way the light reaches us. Consider a distant metre stick moving with speed v at right angles to your direction of view, oriented along its direction of motion. Suppose it has thickness ΔL (this thickness being along our line of sight).

a) Draw a diagram to show that you can see the rear edge of the stick.

b) Show further than the angle subtended by the rear edge of the stick is the same angle that it would subtend if it the stick were stationary and rotated by an angle $\theta = \sin^{-1}(v/c)$.

A real moving object therefore appears rotated, as in the sketch below. The upper stick is stationary, while the lower stick is moving sideways with speed v and has its rear edge visible.

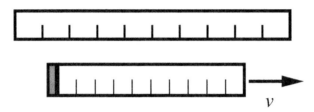

Chapter 3

Tensors in Special Relativity

It is tensors that describe physical quantities · they make physical laws the same for everyone · scalars and 4-vectors are the simplest examples

Doing physics means wanting to know how things behave, how they evolve, and how they influence other things and are in turn influenced by them. These things might be dynamical objects following a path, electric fields pervading a medium, subatomic particles colliding and changing identity, and radiation emerging from an exploding star and traversing the Universe — the possibilities are limited only by physicists' imagining of what can, in principle, be explained. Often, we are interested in how things are changing, whether it be to evolve in time (such as the location of a ball on a slope) or to vary in space (like the gravitational influence of the Sun), and so the natural language to use to construct physical laws describing these variations is that of differential equations.

This chapter examines which types of quantities are useful for assembling physical laws, given that the principle of Special Relativity forces us to consider all inertial observers, and hence coordinate systems, as equally valid describers of events. Hence, the physical quantities have an existence which is independent of the particular set of coordinates being used. If they are specified within a particular coordinate system, they must have a well-defined transformation law that tells us how they change when we switch coordinate systems. Such quantities are known as **tensors**.[1]

In this chapter, we describe some of the properties of tensors within the framework of Special Relativity, developing a notation that can be augmented into full

---

[1] More accurately, it is the quantity at a single location that is a tensor. Usually, there is a tensor at each location in space–time, whose value might vary with location, and this should properly be called a **tensor field**. But commonly 'tensor' is used for both.

Introducing General Relativity, First Edition. Mark Hindmarsh and Andrew Liddle.
© 2022 John Wiley & Sons Ltd. Published 2022 by John Wiley & Sons Ltd.
Companion website: www.wiley.com/go/hindmarsh/introducingGR

General Relativity. In fact, as we will see, many aspects actually finish up looking more straightforward within the general setting, where the transformation terms take the form of the usual chain rule for combining derivatives. But to develop intuition, it is useful to first see how it works for Special Relativity.

3.1 Coordinates

To use tensors effectively, we need to develop an appropriate notation. We begin with the coordinates of an event by writing $x^0 \equiv ct$, $x^1 \equiv x$, $x^2 \equiv y$, and $x^3 \equiv z$. The raised indices are not powers, but are labels indicating which of the time and space coordinates is being considered. We can now denote the coordinates collectively as x^μ, where the index μ takes the values 0, 1, 2, 3.[2] We can write the coordinates of a world line of a particle as a set of four functions $x^\mu(\lambda)$, where λ is a real number labelling points on the line.

As we will be explaining, in tensor algebra applied to Special Relativity, it is vital to distinguish quantities where space–time indices are raised from those where they are lowered. Those quantities are related, but they are not the same. The only exception we have is the labelling of coordinates, where we will *always* adopt raised indices.

With the condensed notation x^μ for the space and time coordinates, we can write the Lorentz transformation as a matrix multiplication of the form

$$x^\mu \to x'^\mu = \sum_{\nu=0}^{3} \Lambda^\mu_{\;\nu} x^\nu \,, \tag{3.1}$$

where

$$\Lambda^\mu_{\;\nu} = \begin{pmatrix} \gamma & -\gamma v/c & 0 & 0 \\ -\gamma v/c & \gamma & 0 & 0 \\ 0 & 0 & 1 & 0 \\ 0 & 0 & 0 & 1 \end{pmatrix} \tag{3.2}$$

for the special case of a transformation between frames moving with respect to one another in the x^1 coordinate direction, with μ labelling the rows of the matrix, and ν the columns. In general, there is a matrix $\Lambda^\mu_{\;\nu}$ for any Lorentz transformation, which could for instance be constructed by multiplying the matrices for transformations along the principal axes. Notice that the lower index has been indented so that it is not directly below the upper index; the horizontal order of indices matters and any tensor must be written in a way that makes the order clear.

---

[2] Some older treatments of Special Relativity define x^0 to be an imaginary number by including an 'i' in its definition. We are not doing that.

Our convention will be to use Greek letters to represent indices running over the values 0, 1, 2, 3, while Latin letters will run over only the spatial coordinates 1, 2, 3. There will be occasional cases where we consider toy example space–times that have fewer than three spatial dimensions, in which case the Greek indices continue to run over all the available values and Latin over the spatial ones. For example, the distance l of a point x from the origin in a frame with space coordinates x^i is given by

$$l^2 = \sum_{i=1}^{3} x^i x^i. \tag{3.3}$$

A very useful convention is the **Einstein summation convention**, in which a repeated index in a term (e.g. a product of tensors) indicates a summation over that index. We will see later that in order to form objects consistent with the postulates of Special Relativity, one space–time index should always be raised and one lowered in the pair of repeated indices. For example

$$\Lambda^{\mu}_{\nu} x^{\nu} \equiv \sum_{\nu=0}^{3} \Lambda^{\mu}_{\nu} x^{\nu}. \tag{3.4}$$

An index which is not repeated, like μ in the above equation, is called free. Every term in a tensor equation must have the same free indices, both in label and position. Each term can have any number of pairs of repeated indices, but the letter representing them must be used only once for each pair to avoid ambiguity. Summation over a repeated index is also called **contraction**.

Index notation can also be adapted to express the space–time interval between points, by introducing a two-index object called a **metric**. Denoting the coordinate differences as Δx^{μ}, we can write

$$\Delta s^2 = -(\Delta x^0)^2 + (\Delta x^1)^2 + (\Delta x^2)^2 + (\Delta x^3)^2, \tag{3.5}$$

and then express it as

$$\Delta s^2 = \sum_{\mu\nu} \eta_{\mu\nu} \Delta x^{\mu} \Delta x^{\nu} \equiv \eta_{\mu\nu} \Delta x^{\mu} \Delta x^{\nu}, \tag{3.6}$$

where, with μ labelling rows and ν the columns.

$$\eta_{\mu\nu} = \begin{pmatrix} -1 & 0 & 0 & 0 \\ 0 & 1 & 0 & 0 \\ 0 & 0 & 1 & 0 \\ 0 & 0 & 0 & 1 \end{pmatrix}_{\mu\nu}. \tag{3.7}$$

Note that it is common in particle physics to define the metric with the opposite sign, so that $\eta_{00} = +1$. This is one of a number of arbitrary choices of sign, known as **sign conventions**, in the General Relativity formalism, which can make comparing results from different sources tricky.

Thus, if we think of Δx^μ as the components of a column vector Δx, we can write the interval as a matrix expression

$$\Delta s^2 = \Delta x^T \eta\, \Delta x. \tag{3.8}$$

The object η is known as the **Minkowski metric** after Hermann Minkowski, who used it to reformulate Special Relativity in a more geometric manner. The components of the Minkowski metric in Cartesian coordinates remain unchanged if we shift between inertial frames. This follows immediately from invariance of the interval, which takes the same form (3.5) in the S' coordinate frame, or can be viewed as a property of Lorentz transformations, as shown in Problem 3.1.

Introducing a 4×4 matrix looks a lot of work for such a simple expression, but there are pay-offs. It allows us to generalise the expression for the space–time interval to any coordinate system. In the general case, the metric components may depend on position, and it makes sense only to consider infinitesimal changes in coordinates dx^μ around a point x. The infinitesimal space–time interval ds is then given by

$$ds^2 = g_{\mu\nu}(x)\, dx^\mu\, dx^\nu. \tag{3.9}$$

Coordinate systems with a position-dependent metric may just be a rewriting of Minkowski space–time, as for example with spherical polar coordinates, or there may be some genuine departure from the Minkowski space–time: understanding and quantifying these differences is the business of General Relativity.

Note finally that the transformation law for differentials dx^μ in Minkowski space–time is the same as the transformation law for the coordinates,

$$dx^\mu \to dx'^\mu = \Lambda^\mu_\nu\, dx^\nu. \tag{3.10}$$

3.2   4-vectors

Now that we have discussed coordinates and the transformation between coordinate systems, we can consider physical quantities that we can describe using those coordinates. We begin with the concept of a 4-vector. At a given point in space–time, a 4-vector $\vec{a}$ can be thought of as an arrow with a length and pointing in some direction *in space–time*. Within a given coordinate system, its projection in each coordinate direction gives its component in that direction, just as in standard

(Euclidean) vector geometry. So it has four components in all, that we write a^μ. An example of a 4-vector is the tangent to a world-line $t^\mu = dx^\mu/d\lambda$. Given that the coordinates of every point on the curve transform according to equation (3.4), the tangent vector transforms in the same way.

We make the definition that any four-component object a^μ which behaves like the differential dx^μ under coordinate transformations is known as a **contravariant 4-vector**

$$a^\mu \to a'^\mu = \Lambda^\mu_\nu a^\nu . \tag{3.11}$$

This is exactly the same as equation (3.10) except that the vector a^μ appears in place of the differentials dx^μ. Not all four-component objects are 4-vectors; to be a 4-vector, the object has to have the transformation law (3.11).

We can recover the original 4-vector by a Lorentz boost with the opposite velocity, which defines an inverse Lorentz transformation

$$a'^\mu \to (\Lambda^{-1})^\mu_\nu a'^\nu = (\Lambda^{-1})^\mu_\nu \Lambda^\nu_\rho a^\rho . \tag{3.12}$$

The mathematical definition of the inverse is through

$$(\Lambda^{-1})^\mu_\nu \Lambda^\nu_\rho = \delta^\mu_\rho, \tag{3.13}$$

where δ^μ_ρ is the **Kronecker delta**, satisfying

$$\delta^\mu_\rho = \begin{cases} 1, & \text{if } \mu = \rho, \\ 0, & \text{otherwise.} \end{cases} \tag{3.14}$$

Hence,

$$a'^\mu \to \delta^\mu_\rho a^\rho = \delta^\mu_0 a^0 + \delta^\mu_1 a^1 + \delta^\mu_2 a^2 + \delta^\mu_3 a^3 = a^\mu, \tag{3.15}$$

where we have written out the sum over the index ρ. If you are unused to 4-vector notation, it is worth checking that the final equality is true for $\mu = 0, 1, 2, 3$ explicitly, using the definition of the Kronecker delta.

We can form a scalar product between 4-vectors by using the metric; we define

$$\vec{a} \cdot \vec{b} \equiv a^\mu \eta_{\mu\nu} b^\nu = -a^0 b^0 + a^1 b^1 + a^2 b^2 + a^3 b^3 . \tag{3.16}$$

Such a product is Lorentz invariant, by the same argument that shows that the interval is invariant under Lorentz transformations (see Problem 3.1). That means that all inertial observers agree on the value of the product, even though they don't agree on the components of the individual vectors being multiplied. The product is a statement about the vectors which does not depend on the coordinates being used.

When writing a 4-vector, or a tensor more generally, it matters whether indices are raised or lowered; they refer to objects which transform in different ways. To demonstrate this, let us introduce the notation

$$b_\mu \equiv \eta_{\mu\nu} b^\nu, \tag{3.17}$$

which implies $b_0 = -b^0$ and $b_i = b^i$. Under a Lorentz transformation,

$$a^\mu b_\mu \rightarrow a'^\mu b'_\mu = a^\mu b_\mu. \tag{3.18}$$

The invariance of the scalar product implies that b_μ must be boosted with the inverse Lorentz transformation

$$b_\mu \rightarrow b'_\mu = (\Lambda^{-1})^\rho{}_\mu b_\rho. \tag{3.19}$$

A four-component object transforming with the inverse is known as a **covariant 4-vector** and written with a lowered index. Clearly, it is closely related to its contravariant cousin, from which it is obtained by multiplication with the metric according to equation (3.17), a procedure known as lowering the index.

Note that in the expression for the scalar product, we could also have chosen to lower the index on the other 4-vector,

$$a^\mu \eta_{\mu\nu} b^\nu = a^\mu b_\mu = a_\nu b^\nu = a_\mu b^\mu, \tag{3.20}$$

where we have used the freedom to relabel summed indices in the last equality. Hence in a summed pair of indices, we may switch the raised and lowered indices, $a^\mu b_\mu = a_\mu b^\mu$.

A common notation when taking the scalar product of a 4-vector with itself is

$$\vec{a} \cdot \vec{a} \equiv a^2. \tag{3.21}$$

This applies equally to contravariant and covariant vectors. There is the potential for confusion between a^2 and one of the spatial components of the 4-vector $\vec{a}$, but the meaning should be clear from the context. In particular, we should not expect to see one of the spatial components singled out in the formulation of a fundamental physical law.

In forming the scalar product, we sum over one raised index and one lowered index. The result is invariant under a Lorentz transformation. The Einstein summation convention can in principle be applied to a pair of indices which are both raised or both lowered, such as $a^\mu a^\mu$, but this expression, equal to $(a^0)^2 + (a^1)^2 + (a^2)^2 + (a^3)^2$, is not invariant under Lorentz transformations like the scalar product. It should therefore not appear in any equations describing physical laws which are supposed to be consistent with Special Relativity. However, one should not be surprised to see expressions like $a^i a^i$, which is just the modulus-squared of the spatial parts of a 4-vector in a particular frame.

We can also turn a covariant 4-vector into a contravariant one, using the inverse metric, defined by

$$\eta^{\mu\nu}\eta_{\nu\rho} = \delta^\mu_\rho. \tag{3.22}$$

The inverse metric is written with both its indices raised. It can be used to raise indices on 4-vectors, and hence obtaining a covariant 4-vector, according to

$$b^\mu = \eta^{\mu\nu}b_\nu. \tag{3.23}$$

We can check that this definition is consistent by lowering and then raising again, checking that we get the same answer

$$\eta^{\mu\nu}(\eta_{\nu\rho}b^\rho) = \delta^\mu_\rho b^\rho = b^\mu, \tag{3.24}$$

with the multiplication by the Kronecker delta being familiar from equation (3.15).

The process of raising and lowering indices can be applied to the Lorentz transformation,

$$b_\mu \to b'_\mu = \eta_{\mu\nu}b'^\nu = \eta_{\mu\nu}\Lambda^\nu{}_\rho b^\rho = \Lambda_{\mu\rho}b^\rho = \Lambda_\mu{}^\rho b_\rho. \tag{3.25}$$

This superficially looks the same as the transformation of a contravariant 4-vector, but the order of the indices matters. Hence, we conclude that

$$\Lambda_\mu{}^\rho = (\Lambda^{-1})^\rho{}_\mu, \tag{3.26}$$

that is, the Lorentz transformation written with the indices in reverse order is the inverse (see also Problem 3.6).

We can also think of $\Lambda^\mu{}_\nu$ and $\eta_{\mu\nu}$ as matrices Λ and η, with the first index labelling rows and the second columns. Exchanging indices is equivalent to taking the transpose of the matrix, and so we can write equation (3.26) as $\eta^{-1}\Lambda^T\eta = \Lambda^{-1}$, where η^{-1} is the inverse metric. Using the fact the components of η^{-1} are the same as those of η, we can write the matrix equation

$$\eta\Lambda^T\eta = \Lambda^{-1}. \tag{3.27}$$

Thus, the inverse Lorentz transform is (after conjugation by η) the transpose. From this equation, one can glimpse the deep connection of Lorentz transformations to orthogonal transformations (i.e. rotations).

Indeed, the definition of a Lorentz transformation as the transformation on a 4-vector which preserves the scalar product and possesses an inverse is enough to define a group of transformations, the **Lorentz group**, which includes ordinary three-dimensional rotations as well as Lorentz boosts (see Problem 3.7).

To end this discussion, we wish to make it clear that there is a distinction between the 4-vector itself, $\vec{a}$, and its components a^μ. The vector exists independently of the coordinate system. The same concept exists for 3-vectors, where we can write, say, the velocity 3-vector as $\mathbf{v}$, and its components with respect to a set of three basis vectors $\mathbf{e}_i$, such that $\mathbf{v} = v^i \mathbf{e}_i$. The vector $\mathbf{v}$ does not change under a change of basis $\mathbf{e}_i \to \mathbf{e}'_i$, only its components do. Similarly, 4-vector components are related to the 4-vector through a set of four space–time vectors $\vec{e}_\mu$, so that $\vec{a} = a^\mu \vec{e}_\mu$. In this book, we will use the common, but inaccurate, convention of referring to vectors and tensors by their components; for example, we will refer to the *vector a^μ* rather than the *components of the vector a^μ*.[3]

3.3   4-velocity, 4-momentum, and 4-acceleration

We now consider motions of objects in space–time, bearing in mind that we need to ensure we retain well-defined transformation laws. To generate a velocity which will transform as a 4-vector, we cannot differentiate with respect to time, because each observer has a different opinion on what time is. But fortunately, there is a time that everyone can agree on, which is **proper time** defined by

$$d\tau^2 = -\frac{ds^2}{c^2}, \tag{3.28}$$

where $ds^2 = -dx^\mu dx_\mu$ is the interval, which is Lorentz invariant. Any massive object is restricted to stay within its forward light cone, making ds^2 negative under our sign conventions; we include a minus sign in the definition and take the positive square root so that proper time increases towards the future.

Now, we can define the 4-velocity of the object by

$$u^\mu = \frac{dx^\mu}{d\tau}. \tag{3.29}$$

---

[3]Our introduction of the transformation law for vectors in terms of partial derivatives of coordinates is specific to a particular class of bases, known as coordinate bases. One can generalise the discussion to non-coordinate bases, at the expense of extra technical complexity. The book by Bernard Schutz, listed in the bibliography, has a thorough and pedagogical discussion on both types of basis.

Since τ is Lorentz invariant, the 4-velocity inherits the correct transformation law for a contravariant 4-vector directly from the coordinates. Note that we cannot use proper time in the case of a particle like a photon moving at the speed of light, as $ds = 0$ along its trajectory in space–time. We discuss photon 4-velocities at the end of this section.

Proper time τ has the important property that it is the time recorded by a clock moving with the object whose 4-velocity is u^μ. If you consider the 'clock' to be your beating heart, you will realise that proper time is the time experienced by any observer as they follow their world line.

This can be shown by transforming so that the observer is at rest as follows. Suppose the observer has 3-velocity $v^i = dx^i/dt$ in some frame. Then the 4-velocity is

$$u^\mu = \left(c\frac{dt}{d\tau}, \frac{dt}{d\tau}\frac{dx^i}{dt} \right). \tag{3.30}$$

Boost to a new frame moving with velocity v^i, so that the observer is at rest. In the new frame, the 4-velocity is

$$u'^\mu = \left(c\frac{dt'}{d\tau}, 0 \right). \tag{3.31}$$

In this frame, infinitesimal changes in the coordinates along the world line are entirely in the time direction, and so $d\tau = dt'$. Hence, τ is indeed time as experienced by this observer.

We also know that $dt = \gamma dt' = \gamma d\tau$ (the observer is not moving in their rest frame, so $dx' = 0$). Hence, $dt/d\tau = \gamma$, and in the original frame we can write the 4-velocity as

$$u^\mu = \left(\gamma c, \gamma\frac{dx^i}{dt} \right). \tag{3.32}$$

In particular, for a stationary object, the components of the 4-velocity are given by $u^\mu = (c, 0, 0, 0)$. From this, it is immediately clear that

$$u_\mu u^\mu = -c^2. \tag{3.33}$$

As we have taken the scalar product, this equation is true in all frames. We can interpret it as a statement that objects all move at the speed of light through space–time. It can also be shown directly by squaring equations (3.29) or (3.32).

From the 4-velocity, we can define the 4-momentum

$$p^\mu = m_0 u^\mu, \tag{3.34}$$

where m_0 is the rest-mass (i.e. the mass in the frame where the object is stationary). Again, the correct transformation law is inherited from the velocity. Note that if we want to define the relationship between the 3-velocity and the 3-momentum, using equation (3.32), we would get $p^i = m_0 \gamma v^i$. We can interpret this as the particle gaining a relativistic mass $m = m_0 / \sqrt{1 - v^2/c^2}$. In that sense moving objects appear to be heavier, even though it is just the rest mass that appears in the relativistic formula for the 4-momentum.

A 4-vector acceleration can be defined by differentiating the 4-velocity with respect to proper time,

$$a^\mu \equiv \frac{du^\mu}{d\tau} = \frac{d^2 x^\mu}{d\tau^2} . \tag{3.35}$$

A free particle moves with zero 4-acceleration.

Particles with zero rest mass, like photons, move on paths in space–time for which $ds = 0$, and so their 4-velocity can't be defined as the rate of change of position with respect to proper time. The path can still be parametrised by a real number λ, but it has no particular physical interpretation, and the path $x^\mu(\lambda)$ must satisfy

$$u_\mu u^\mu = \frac{dx^\mu}{d\lambda} \frac{dx_\mu}{d\lambda} = 0 . \tag{3.36}$$

Any other parameter related by a linear transformation $\lambda' = a\lambda + b$ will also be acceptable, as linear transformations maintain the vanishing of the 4-acceleration. For example, we can describe a photon moving in the z direction by $x^3 = \lambda$, $x^0 = c\lambda$.

3.4 4-divergence and the wave operator

We now come to the key issue of differentiation. The most common forms of physical law express rates of change with respect to the coordinates, not just to proper time. We can't pick out a single coordinate direction and consider it in isolation, because the Lorentz transformation mixes coordinates when we shift frame. Instead, we must consider the coordinate derivatives collectively:

$$\frac{\partial}{\partial x^\mu} \equiv \left(\frac{\partial}{\partial x^0}, \frac{\partial}{\partial x^1}, \frac{\partial}{\partial x^2}, \frac{\partial}{\partial x^3} \right) . \tag{3.37}$$

This looks like it ought to be a 4-vector, but what kind is it?[4] The transformation law comes from the chain rule

$$\frac{\partial}{\partial x^{\mu}} \rightarrow \frac{\partial}{\partial x'^{\mu}} = \frac{\partial x^{\nu}}{\partial x'^{\mu}} \frac{\partial}{\partial x^{\nu}}. \tag{3.38}$$

Now,

$$x'^{\mu} = \Lambda^{\mu}_{\ \nu} x^{\nu}; \quad x^{\mu} = (\Lambda^{-1})^{\mu}_{\ \nu} x'^{\nu}. \tag{3.39}$$

Differentiating both sides of the latter with respect to x'^{ρ} gives

$$\frac{\partial x^{\mu}}{\partial x'^{\rho}} = (\Lambda^{-1})^{\mu}_{\ \nu} \frac{\partial x'^{\nu}}{\partial x'^{\rho}}, \tag{3.40}$$

where the last term is just a Kronecker delta δ^{ν}_{ρ}. So

$$\frac{\partial}{\partial x^{\mu}} \rightarrow \frac{\partial}{\partial x'^{\mu}} = (\Lambda^{-1})^{\nu}_{\ \mu} \frac{\partial}{\partial x^{\nu}}. \tag{3.41}$$

So the derivative $\partial/\partial x^{\mu}$ transforms as a covariant 4-vector.

This is a key result. The transformation law of a derivative is *not* the same as the transformation law of the coordinates. This helps us understand why we have to deal with two types of 4-vectors in order to write differential equations in tensor form. In considering 4-vectors, we will want to know whether they are of a type which transforms as the coordinates (contravariant, raised index), or as the derivative with respect to those coordinates (covariant, lowered index).

Because the derivative transforms as a covariant vector, it is often abbreviated to $\partial/\partial x^{\mu} \equiv \partial_{\mu}$.

A particularly common differential operator appearing in physical laws is the wave operator (the D'Alembertian), comprised of the sum of the second derivatives. It is defined as

$$\Box = -\frac{1}{c^2} \frac{\partial^2}{\partial t^2} + \frac{\partial^2}{\partial (x^1)^2} + \frac{\partial^2}{\partial (x^2)^2} + \frac{\partial^2}{\partial (x^3)^2}, \tag{3.42}$$

and can be written extremely compactly as

$$\Box \equiv \eta^{\mu\nu} \partial_{\mu} \partial_{\nu} \equiv \partial^{\mu} \partial_{\mu}. \tag{3.43}$$

---

[4]The derivative is here written as an operator, i.e. to have meaning in an equation, it has to act upon some quantity. But it still makes sense to ask what transformation property the differential operator satisfies.

The wave operator is invariant under Lorentz transformations. This is the feature of electromagnetism which ensures that the speed of light is the same for all inertial observers.

3.5   Tensors

Tensors generalise the idea of 4-vectors to arbitrary numbers of indices. In essence they are multi-dimensional arrays which have specific transformation properties. An example tensor is the product of two vectors

$$M^{\mu\nu} \equiv a^\mu b^\nu \,, \tag{3.44}$$

which will transform according to

$$M^{\mu\nu} \to M'^{\mu\nu} = a'^\mu b'^\nu = \Lambda^\mu{}_\rho a^\rho \Lambda^\nu{}_\sigma b^\sigma = \Lambda^\mu{}_\rho \Lambda^\nu{}_\sigma M^{\rho\sigma} \,. \tag{3.45}$$

Note there is a transformation matrix for each index. The metric tensor is an example of a tensor with two indices.

Not all two-index tensors can be written as products of 4-vectors. More generally, the transformation law above defines what is meant by such a tensor.

The number of indices is known as the **rank** of a tensor. A 4-vector is a particular case of a tensor, a rank-1 tensor. Tensors can be contravariant (all indices raised), covariant (all indices lowered) or mixed. If we wish to specify the number of covariant and contravariant indices, we use the notation $\binom{m}{n}$ to denote a tensor with m contravariant and n contravariant indices. For example, the metric $\eta_{\mu\nu}$ is a rank-$\binom{0}{2}$ tensor. Indices are raised and lowered using the metric tensor. In this book, we will see tensors of various ranks, the highest being the rank-4 Riemann tensor needed to describe curvature of space–time.

The special case of a rank-0 tensor is known as a **scalar**. It has no indices and so needs no transformation term when we move between frames. Hence, a scalar has the same value, at a given space–time point, in any coordinate system. Examples we have already seen are the interval and the dot product of two 4-vectors. Be sure to realise that the value of a scalar quantity can still vary in time and space; the invariance refers to its value *at a given space–time point as measured in different coordinate systems.*

Various processes can change the rank of a tensor, for example multiplication, differentiation, and contraction which are illustrated by the following three examples

$$
\begin{aligned}
c_{\mu\nu} &= a_\mu b_\nu \,; \\
c_\mu{}^\nu &= \partial_\mu a^\nu \,; \\
a_\mu &= c_{\mu\nu}{}^\nu \,.
\end{aligned}
\qquad (3.46)
$$

The last of these, contraction, refers to summing over pairs of indices in a tensor. Because one such index is always contravariant and the other covariant, they have opposite transformations which then cancel out, leaving the simpler transformation law of a tensor whose rank is smaller by 2. Hence, the last equation above has tensor rank 1; even though the right-hand side superficially looks like it might have rank 3, it is the number of free (i.e. unsummed) indices that determines the rank.

Our purpose in introducing tensors is to use them to describe physical laws, i.e. they are to appear in equations with the goal that these equations take the same form independent of the coordinates being used. This will guarantee that the laws of physics are the same in all inertial frames. To achieve this, we have to ensure that each side of our equations has the same transformation law. This is guaranteed if each term in the equations has the same tensor rank and indexing scheme for any free indices. Hence, the rules of assembling tensor equations are simply that

1. Any free indices must match up in the different terms, and be at the same level. That is, all terms must be tensors of the same rank.

2. Any pair of repeated indices in a given term are summed over.

3. Pairs of repeated indices must be at different levels.

Together these rules ensure that the terms in our equations match properly both in terms of the number of components and the transformation law of each, making sure for instance that we are not attempting something nonsensical like equating a scalar to a 4-vector.

The actual symbols used for indices are irrelevant and can be changed at will; this often has to be done during calculations. It is important not to use the same symbol for the index in two different sums, which makes expressions ambiguous and/or inconsistent; if the same index symbol appears more than twice in any expression (expression here typically meaning a product of various tensors and their derivatives), or more than once in either the raised or lowered position, you have almost certainly duplicated a symbol and gone wrong.

The beauty of a tensor equation is the automatic way that it is guaranteed to be unchanged by coordinate transformation. Each term in the equation will have some free indices, and so will acquire some Lorentz transformation matrices when we change frame. But every term has the same indices and hence ends up with exactly the same transformation matrices. So at the end, we arrive back to the same tensor equation we started with, now applying in the new coordinate frame. Hence, tensors are the natural language for expressing physical laws in a coordinate-independent fashion.

3.6   Tensors in action: the Lorentz force

In Section 3.5, we stated that physical laws formulated as tensor equations are guaranteed to have the same form in all inertial frames. In this section, we will give an example from electromagnetism, the Lorentz force law.

We first examine how Newton's second law is written in a Lorentz-covariant manner. We have already seen how to write the acceleration in a covariant way, as the derivative of the 4-velocity with respect to proper time, so there must also be a 4-vector force f^μ, such that

$$f^\mu = m_0 \frac{du^\mu}{d\tau}.$$ (3.47)

It may look as if Special Relativity is demanding four equations rather than the three given by Newton, but in fact, the 0th component of this equation is not independent. Multiplying both sides by the 4-velocity and contracting the index we find

$$u_\mu f^\mu = m_0 u_\mu \frac{du^\mu}{d\tau} = \frac{1}{2} m_0 \frac{du^2}{d\tau} = 0.$$ (3.48)

The last equality follows because $u^2 = -c^2$, a constant, as we saw in equation (3.33). Hence, the force 4-vector must be orthogonal (in the space–time sense) to the 4-velocity.

Recalling that the 0th component of the momentum is the energy E, we see that the 0th component of the 4-force is the rate of doing work on the particle. Hence, the orthogonality condition of equation (3.48), which may be written $\gamma(-f^0 + f^i v^i) = 0$, is just an expression of the fact that the rate of doing work is the scalar product of the 3-vector force and the velocity.

In electromagnetism, the 3-vector force $\mathbf{F}$ on a particle with charge q moving with velocity $\mathbf{v}$ is given by the Lorentz force, $\mathbf{F} = q(\mathbf{E} + \mathbf{v} \times \mathbf{B})$. In a Cartesian basis, the equations for the components of the force becomes

$$F^i = q\left(E^i + \varepsilon^{ijk} v^j B^k\right),$$ (3.49)

where we introduce the **Levi-Civita symbol** ε^{ijk}, named after Tullio Levi-Civita and defined by

$$\varepsilon^{ijk} = \begin{cases} +1 & \text{if } ijk \text{ is an even permutation of } 123, \\ -1 & \text{if } ijk \text{ is an odd permutation of } 123, \\ 0 & \text{otherwise.} \end{cases} \tag{3.50}$$

Note that for 3-vectors, there is no difference between up and down indices in Cartesian coordinates.

The electric and magnetic fields E^i and B^i belong together in a Lorentz-covariant object, as they are mixed by Lorentz transformations. However, this object cannot be a 4-vector, as there are a total of six components of the electric and magnetic fields when they are taken together. In fact, the object is a rank-2 tensor, the field-strength tensor $F^{\mu\nu}$, whose entries are

$$F^{\mu\nu} = \begin{pmatrix} 0 & E^1/c & E^2/c & E^3/c \\ -E^1/c & 0 & -B^3 & B^2 \\ -E^2/c & B^3 & 0 & -B^1 \\ -E^3/c & -B^2 & B^1 & 0 \end{pmatrix}, \tag{3.51}$$

where μ labels rows and ν columns. There is an equivalent expression in terms of indices, $F^{0j} = E^j/c$ and $F^{ij} = \varepsilon^{ijk}B^k$.

Note that the field-strength tensor is *antisymmetric*, i.e.

$$F^{\mu\nu} = -F^{\nu\mu}. \tag{3.52}$$

One can check (see Problem 3.12) that the standard rank-2 tensor Lorentz transformation rule (3.45) gives the correct transformation law for the electric and magnetic fields.

Expressed in terms of the field-strength tensor, the Lorentz force 4-vector is

$$f^\mu = -qF^{\mu\nu}u_\nu, \tag{3.53}$$

a significantly more compact expression than the 3-vector version. The orthogonality requirement $u_\mu f^\mu = 0$ can be seen to be satisfied by an instructive index relabelling trick (see Problem 3.10). Maxwell's equations for the electric and magnetic fields also have an elegant form in relativistic notation (see Problem 3.11).

Problems

3.1* In tensor notation the Lorentz transformation can be written as

$$x^\rho \to x'^\rho = \sum_{\sigma=0}^{3} \Lambda^\rho{}_\sigma x^\sigma .$$

The space–time interval can be written as

$$ds^2 = \sum_{\mu\nu} \eta_{\mu\nu} \, dx^\mu \, dx^\nu .$$

a) Show that invariance of the interval under Lorentz transformations implies that

$$\sum_{\mu\nu} \eta_{\mu\nu} \Lambda^\mu{}_\rho \Lambda^\nu{}_\sigma = \eta_{\rho\sigma} .$$

b) Show that this equation may be written in matrix form

$$\Lambda^T \eta \Lambda = \eta .$$

c) Demonstrate the Lorentz invariance of the scalar product between two contravariant 4-vectors a and b written as column vectors, $a \cdot b = a^T \eta b$.

3.2 Equation (3.32) shows that the 4-velocity of an object moving with 3-velocity dx^i/dt is

$$u^\mu = \left(\gamma c, \gamma \frac{dx^i}{dt} \right) .$$

Demonstrate that $u_\mu u^\mu = -c^2$.

The 4-momentum of a moving object is defined using its rest mass m_0 as $p^\mu = m_0 u^\mu$. If we define mass using the 3-momentum and 3-velocity, so that $p^i = m \, dx^i/dt$, how is m related to m_0?

3.3 Write out in component form the vector equation

$$a'^\mu = M^\mu{}_\nu a^\nu ,$$

without making any assumption as to the form of $M^\mu{}_\nu$. If $M^\mu{}_\nu = \delta^\mu{}_\nu$, the Kronecker delta, show that $a'^\mu = a^\mu$.

3.4* Taking $\eta_{\mu\nu}$ to be the Minkowski metric,

    a) Evaluate $\eta_{\mu\nu}\eta^{\mu\nu}$.

    b) Simplify as far as possible the expression $\eta_{\mu\nu}\eta^{\mu\rho}\eta^{\nu\sigma}$.

3.5   Write down the transformation law for a three-index tensor $M^{\mu}_{\ \nu\rho}$ under a Lorentz transformation.

3.6   The inverse Lorentz transformation is written $(\Lambda^{-1})^{\mu}_{\ \nu}$, defined so that

$$(\Lambda^{-1})^{\mu}_{\ \nu}\Lambda^{\nu}_{\ \rho} = \delta^{\mu}_{\rho}.$$

    a) Considering the special case of a Lorentz boost in the x-direction, verify that the inverse is obtained by substituting $v \to -v$.

    b) The forward Lorentz transformation is $x'^{\mu} = \Lambda^{\mu}_{\ \nu}x^{\nu}$. Show that

$$x^{\mu} = (\Lambda^{-1})^{\mu}_{\ \nu}x'^{\nu}.$$

    c) In Problem 3.1, you showed that a general Lorentz transformation satisfies $\eta_{\mu\nu}\Lambda^{\mu}_{\ \rho}\Lambda^{\nu}_{\ \sigma} = \eta_{\rho\sigma}$. Use this equation to show that

$$(\Lambda^{-1})^{\mu}_{\ \nu} = \Lambda_{\nu}^{\ \mu}.$$

3.7   Suppose that a Lorentz transformation matrix has the form

$$\Lambda = \begin{pmatrix} 1 & \mathbf{0}^T \\ \mathbf{0} & \Omega \end{pmatrix},$$

where Ω is a 3×3 matrix, and $\mathbf{0}$ is a column vector of three zeros. Show from equation (3.27) that Ω is an orthogonal matrix. This demonstrates that a rotation is also a Lorentz transformation.

3.8   State whether the following are satisfactory tensor equations. If they are not, state why. If they are, state the tensor rank of the equation.

    a) $\partial_{\mu}a_{\nu} = M_{\mu\rho}$.

    b) $a_{\mu}a^{\mu} = \partial_{\nu}a^{\nu}$.

    c) $\partial_{\rho}M^{\rho\sigma}N_{\sigma}^{\ \nu}P^{\rho}_{\ \lambda} = \partial^{\nu}A$.

    d) $\partial_{\nu}F^{\mu\nu} = j^{\mu}$.

e) $R_{\mu\nu} = R^{\lambda}{}_{\mu\lambda\nu}$.

f) $R_{\mu\nu\rho\sigma} + R_{\mu\rho\sigma\nu} + R_{\mu\sigma\nu\rho} = 0$.

3.9* Given that the derivative ∂_{μ} transforms as a covariant vector, show that ∂^{μ} transforms as a contravariant vector.

3.10* Show that the Lorentz 4-force f^{μ} of equation (3.53) is space–time orthogonal to the 4-velocity u^{μ}, i.e. that $u_{\mu}f^{\mu} = 0$.

3.11 [Hard!] Maxwell's equations for electromagnetism in vacuum may be written

$$\partial_i E^i = \frac{1}{\varepsilon_0}\rho_{\mathrm{e}}, \qquad \varepsilon^{ijk}\partial_j B^k = \mu_0 J^i + \mu_0\varepsilon_0 \dot{E}^i,$$

$$\partial_i B^i = 0, \qquad \varepsilon^{ijk}\partial_j E^k = -\dot{B}^i.$$

where ρ_{e} is the electric charge density, J^i are the components of the electric current, and ε_0 and μ_0 are the permittivity and permeability of free space, whose product determines the speed of light via $\varepsilon_0\mu_0 = 1/c^2$.

a) Show that the first two may be combined into one relativistic equation

$$\partial_\nu F^{\mu\nu} = j^\nu,$$

and find the expression for the 4-vector current j^ν.

b) Introducing the 4-dimensional Levi-Civita symbol $\varepsilon_{\mu\nu\rho\sigma}$, defined by

$$\varepsilon_{\mu\nu\rho\sigma} = \begin{cases} +1 & \text{if } \mu\nu\rho\sigma \text{ is an even permutation of 0123,} \\ -1 & \text{if } \mu\nu\rho\sigma \text{ is an odd permutation of 0123,} \\ 0 & \text{otherwise.} \end{cases}$$

show that the other two Maxwell equations may be written as

$$\varepsilon_{\mu\nu\rho\sigma}\partial^\nu F^{\rho\sigma} = 0.$$

3.12* Show that the standard rank-2 tensor Lorentz transformation rule (3.45) gives the correct transformation law for the electric and magnetic fields in the case of a frame moving in the $+x$ direction with speed v:

$$
\begin{array}{llll}
E^1 \rightarrow E'^1 &=& E^1 & \qquad B^1 \rightarrow B'^1 &=& B^1 \\
E^2 \rightarrow E'^2 &=& \gamma(E^2 + vB^3) \ ; & \qquad B^2 \rightarrow B'^2 &=& \gamma(B^2 - vE^3/c^2) \\
E^3 \rightarrow E'^3 &=& \gamma(E^3 - vB^2) & \qquad B^3 \rightarrow B'^3 &=& \gamma(B^3 + vE^2/c^2)
\end{array} \ .
$$

Chapter 4

Towards General Relativity

*Gravity as Newton saw it · can we make Newtonian gravity
relativistic? · the Equivalence Principle shows us the way · light
bends and redshifts · space–time must be curved!*

4.1 Newtonian gravity

For over 200 years until Einstein's innovative discoveries, Isaac Newton's theory of gravity was the established description of gravitational forces, providing a
beautiful unification of the everyday force we experience on Earth with the force
responsible for binding together astronomical objects. In conjunction with the
laws of motion, Newtonian dynamics was a complete description of the move-
ments of planets, stars, and galaxies, and is still accurate enough to be used for
the navigation of space probes around the Solar System.

Newton expressed his theory as a force law, the gravitational force exerted by
a body of mass m_1 on another of mass m_2 being

$$\mathbf{F}_{12} = -G\frac{m_1 m_2}{r^2}\hat{\mathbf{r}}_{12}\,, \tag{4.1}$$

where r is the distance between the two bodies and $\hat{\mathbf{r}}_{12}$ the unit vector in the
direction from the mass m_1 to the mass m_2. The minus sign indicates the force is
directed so as to reduce that distance. The law is an inverse-square law, i.e. the
magnitude of the force decreases as the square of the separation between the two
objects, while being proportional to each of the masses.

The quantity G is **Newton's gravitational constant**, a fundamental constant
which measures the strength of gravity. This is not specified by theoretical prin-

Introducing General Relativity, First Edition. Mark Hindmarsh and Andrew Liddle.
© 2022 John Wiley & Sons Ltd. Published 2022 by John Wiley & Sons Ltd.
Companion website: www.wiley.com/go/hindmarsh/introducingGR

ciples, but rather has to be measured; its modern value and uncertainty is

$$G = (6.674\,08 \pm 0.000\,31) \times 10^{-11}\,\mathrm{m^3\,kg^{-1}\,s^{-2}}.\qquad(4.2)$$

The surprisingly large uncertainty compared with other fundamental constants is a reflection of the relative weakness of the gravitational force, and of the impossibility of isolating the experimental apparatus from other gravitating bodies.

The effect of the force is an acceleration, according to Newton's Second Law of Motion, with a constant of proportionality which is also a mass,

$$\mathbf{F} = m_i \frac{d\mathbf{v}}{dt}.\qquad(4.3)$$

The subscript here stands for 'inertial', to distinguish it from the mass appearing in the force law (4.1), which we refer to as the gravitational mass (with subscript g) when we need to distinguish the two.

As the gravitational force is conservative, one can introduce a **gravitational potential** Φ, in terms of which the force is

$$\mathbf{F} = -m_g \nabla \Phi.\qquad(4.4)$$

The gravitational potential about a point mass M_g is given by

$$\Phi = -\frac{GM_g}{r},\qquad(4.5)$$

and measures the gravitational potential energy of a unit mass at distance r. The acceleration of a body in a gravitational potential is therefore

$$\frac{d\mathbf{v}}{dt} = -\frac{m_g}{m_i} \nabla \Phi.\qquad(4.6)$$

The observation, originally by Galileo Galilei, that the acceleration of a body in the Earth's gravitational field is independent of its mass motivated Newton to propose that the inertial and gravitational masses of an object are in fact equal for all bodies,

$$m_i = m_g.\qquad(4.7)$$

This assertion is one expression of what is now known as the **Weak Equivalence Principle**, which strictly applies only to test bodies, i.e. those with masses small enough not to significantly disturb the prevailing gravitational field. The stronger statement of the equality of gravitational and inertial masses for all bodies, including their effect on the gravitational field, and the contribution of the field to the masses, is known as the Strong Equivalence Principle. The Weak Equivalence Principle has been tested to about 1 part in 10^{13} by torsion balance experiments, and preliminary results (possibly superseded by the time you are reading this) from the MICROSCOPE satellite experiment have recently improved this by an order of magnitude. In Newtonian mechanics, the equality of inertial and gravitational mass is rather mysterious; in General Relativity, as we will see, it has a geometric origin.

Newtonian gravity is linear, meaning that the gravitational force and potential due to multiple objects is additive. For example, the potential at point **x** due to objects of masses M_n at positions $\mathbf{x}_n$ is

$$\Phi = -\sum_n \frac{GM_n}{|\mathbf{x} - \mathbf{x}_n|} . \tag{4.8}$$

In the limit of a very large number of masses in any volume of interest, one can replace the point particles and their positions with the mass density $\rho(\mathbf{x})$, and the gravitational potential then obeys Poisson's equation

$$\nabla^2 \Phi = 4\pi G \rho . \tag{4.9}$$

At the beginning of the 20th century, Newtonian gravity was still extraordinarily successful. It explained the motion of all the bodies in the Solar System to the precision available, with the exception of the innermost planet Mercury. There the deviation was slight; its perihelion was observed to advance by 43 arcseconds per century more than the detailed application of Newtonian mechanics predicted, with an uncertainty of about 2%. It was far from clear that a new physical theory was needed; most explanations at the time invoked some kind of unseen object or matter perturbing Mercury's orbit.

4.2 Special Relativity and gravity

Given the success of the Special Theory of Relativity in providing a framework for the understanding of electrodynamics, it was natural to seek a theory of gravity consistent with the relativity principle. One would expect an extension of equation (4.9), applying to systems where speeds of bodies are not negligible compared to that of light. The slow speeds of planets in the Solar System would imply that relativistic corrections to their orbits should be small, as the near-perfect accuracy of Newton's theory of gravitation demanded.

The left-hand side of equation (4.9) is easily generalised, as Einstein and others quickly realised, by replacing the Laplacian with the wave operator:

$$\nabla^2\Phi \quad \rightarrow \quad \Box\Phi. \tag{4.10}$$

If Φ is a scalar, then this expression is Lorentz invariant, because the wave operator acting on a tensor produces a tensor of the same rank.

The relativistic generation of Newton's Second Law (4.3) appears straightforward. We already have a 4-vector acceleration, given by equation (3.35), and so it should be equated to a 4-vector gravitational force:

$$F_g^\mu = m_i \frac{du^\mu}{d\tau}. \tag{4.11}$$

Generalising the right-hand side of equation (4.9) and at the same time the left-hand side of equation (4.11) to replace Newton's Second Law is not so easy. The Lorentz-invariant generalisation of mass density should be some kind of density related to the rest masses of the particles involved, but it was not immediately understood how to construct it, or the gravitational force, in a way which conserved energy and momentum.

The first person publish a relativistic theory of gravity which conserved energy and momentum was the Finnish physicist Nordström in 1913. In his theory[1]

$$\Box\Phi = -\frac{4\pi G}{\Phi}T, \tag{4.12}$$

where $T = -c^2\rho + 3p$, and p is the pressure. That the source T is Lorentz invariant follows from its being the trace of a rank-2 tensor, $T = T^\mu{}_\mu$, the energy–momentum tensor discussed in detail in Chapter 6.

The gravitational force in Nordström's theory is

$$F_g^\mu = -\frac{m_i c^2}{\Phi}\left(\partial^\mu\Phi + \frac{u^\mu u^\nu}{c^2}\partial_\nu\Phi\right). \tag{4.13}$$

The second term is required by the condition $F_g^\mu u_\mu = 0$. The effective gravitational mass of a body now depends on the gravitational field; one can write

$$m_g = \frac{m_i c^2}{\Phi}. \tag{4.14}$$

The inertial mass cancels in the dynamical equations for a particle, and so the the-

---

[1]This is actually the second version of his theory, improving the 1912 version after insightful criticisms by Einstein and Max Laue.

ory also obeys the Equivalence Principle. The gravitational mass m_g is constant, and $\Phi/c^2 \to 1$ at infinity so that the inertial and gravitational masses of a particle coincide at long distances. A strange feature is that lengths and times vary in inverse proportion to the gravitational potential; as Einstein and Adriaan Fokker soon discovered, one can interpret this as a space–time interval given by

$$ds^2 = \Phi^2 \left(-c^2 dt^2 + dx^2 + dy^2 + dz^2 \right). \tag{4.15}$$

Nordström's scalar theory was for a while a serious competitor to that of Einstein, who did not work out a satisfactory set of field equations for his more complicated metric theory until 1915. Einstein's objections to it were ideological; he did not like that the inertial mass of a body decreased as it approached others, and objected to the privilege accorded to the coordinates in which the metric is equation (4.15), no matter what the distribution of matter. He also did not like the force law (4.13) because of its feature of producing a component of force in the direction of motion of a body. Nevertheless, the paper with Fokker was important for his thinking about the geometric formulation of the field equations of the metric theory.

Ultimately, the reason to choose Einstein's theory over Nordström's is empirical; they make different physical predictions. Nordström's theory predicts a small perihelion retardation for Mercury (rather than the observed advance), and that light should not be affected by gravitation. The 1919 observation of the deflection of starlight by the Sun was decisive against the scalar theory, at least as the sole source of gravitational forces.

4.3 Motivations for a General Theory of Relativity

We now explore some of the physical motivations that led Einstein towards considering general coordinate transformations. Einstein's thinking has profoundly influenced the way modern theoretical physics is done: we start by deciding which overarching principles we would like our fundamental theory to respect, motivated either by observation or by a theoretical aesthetic. We then seek a mathematical formulation of laws that reflects those principles, hoping that the principles are restrictive enough to leave little or no freedom in how we might do so. The more restrictive our principles, the more predictive we can expect our theory to be. The predictions must, of course, be consistent with experiment and observation.

4.3.1 Mach's Principle

Mach's Principle is the name Einstein gave to Ernst Mach's criticisms of the Newtonian idea of absolute space. Newton justified the idea by a thought experiment with a bucket of water. Imagine setting the bucket rotating. The rotation is communicated to the water through viscosity, and the water climbs up the sides of the bucket. In the frame of the bucket, the water has a flat surface when it starts rotating, and a curved one when it reaches the equilibrium where it appears (in the frame of the bucket) not to be rotating. Only in the frame of reference of absolute space is the water surface flat when not rotating.

Mach believed that we should avoid talking about space, and that it made sense only to talk about relative motion between bodies. He argued that Newton's preferred reference frame must be an effect of the rest of the matter in the Universe. Einstein interpreted Mach's writings to mean that inertia results from the interaction between bodies, and his thinking about the relativity of inertia was an important framework for his ideas about gravity. Ultimately, however, it proved to be rather a dead end, as Mach's Principle is not properly realised in the theory that Einstein came to.

4.3.2 Einstein's Equivalence Principle

Earlier, we discussed the remarkable property that all objects have the same acceleration in a gravitational field. This is a consequence of the equality of inertial and gravitational masses in Newtonian gravity, the Weak Equivalence Principle. Einstein, when developing General Relativity, realised that this property has a geometric origin, and is a consequence of the observation that the effects of gravity can be removed by going into free-fall, provided of course that other forces such as air resistance are absent.

For example, drop an object from height x_0, so $x = x_0 - gt^2/2$, where g is the local acceleration due to gravity. In a frame moving with the body, we have $x' = x - gt^2/2$, and in that frame, the acceleration is

$$a' = \frac{d^2x'}{dt'^2} = \frac{d^2}{dt^2}\left(x - \frac{1}{2}gt^2\right) = a - g = 0. \qquad (4.16)$$

In the free-fall frame, there is no acceleration, and Newton's laws with no gravitational field apply.

Noting that this is a consequence of the Weak Equivalence Principle, Einstein re-formulated it as follows:

> **Einstein's Equivalence Principle:** In a small laboratory which is freely falling in a gravitational field, physical laws are the same as

those observed in an inertial frame without a gravitational field.

To be completely accurate, this is a statement about limits. In any real finite-sized laboratory freely falling in the gravitational field of a real finite-sized object, you can tell that you are in a gravitational field. For example, loose objects within the laboratory will fall towards the centre of the distant gravitating object, and so gradually move together in the frame of the lab. However, this effect can be made as small as you like by making the laboratory smaller and the time of observation shorter. In the limit of an infinitesimally small experiment conducted for an infinitesimally short time, the effect disappears, and this idealised laboratory is said to provide a **local inertial frame**.

As with the postulates of Special Relativity, the Equivalence Principle is taken to be an overarching principle applying to all varieties of physical law. Once you are confined to your idealised freely falling laboratory you can assemble whatever apparatus you like — clocks, lasers, magnets, superconductors, radioactive sources — but as long as you are stuck, there you will not be able to carry out any *local* experiment showing a different result to the one you would get if there were no gravitational field. Therefore, in a local inertial frame, physical laws remain consistent with the more restrictive principle of Special Relativity.

Conversely, it follows that an accelerating reference frame behaves like a gravitational field. To see this, apply an acceleration to the falling laboratory that cancels out the acceleration due to gravity (for example, resting it on the surface of the Earth will do the trick nicely). Apply the same acceleration to a laboratory well away from gravitating objects so that the gravitational field is negligible. We've carried out the same operation on each laboratory, so again the physics within each must be the same. Consider for instance the trajectory of a projectile fired across the laboratory accelerating in space or resting on the Earth (see Figure 4.1). According to the Equivalence Principle, the parabolic trajectory will be just the same. No *local* experiment can distinguish gravity from acceleration.

Contrast then these two views of the gravitational force. According to Newton, when we stand on the Earth's surface, every material particle of the Earth exerts a force on us, through an unspecified but possibly instantaneous mechanism. The effect of this force is balanced by a compensating force from the rigid surface of the Earth, which is transmitted through our shoes and up our bodies and felt by us as the force of gravity. But the Equivalence Principle encourages us to instead think of the freely falling frame which exists locally around us. In that frame, we would feel no gravity, but the surface of the Earth is accelerating *upwards* taking us with it, the acceleration again being transmitted upwards via our shoes. We are all familiar with the feeling; in a lift that is accelerating upwards, that the extra acceleration makes it feel as if gravity were a little bit stronger and we weigh a little bit more. Once the lift stops accelerating, that extra weight disappears, even if the lift continues to ascend.

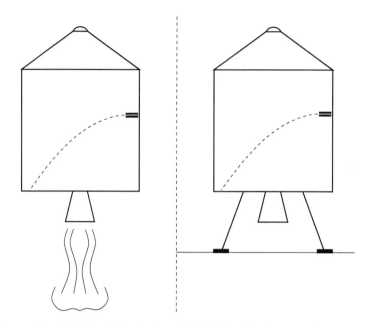

Figure 4.1   The Einstein Equivalence Principle implies that experiments such as this projectile experiment cannot locally distinguish between an accelerating frame (left) and one in a gravitational field (right).

In conclusion, Einstein's formulation of the Weak Equivalence Principle leads us to view gravity as an inertial force. It arises only because we are not using an inertial frame. The freely falling frames are the generalisation of inertial frames in the presence of gravity, and once we use them, we find that local observers cannot distinguish between effects of gravity and effects of acceleration.

4.4   Implications of the Equivalence Principle

The Equivalence Principle matches our experience and hence sounds desirable, but is it actually obeyed in the real Universe? That's something we need to test, and to test it, we need quantitative predictions. In this section, we outline some of the physical consequences which should follow from a theory which obeys the Equivalence Principle. One of the first which Einstein proposed was that particles of light should behave like the projectiles in the example in Section 4.3.2. You can follow the calculation in Problem 4.2. In fact, the effect is more subtle than Einstein originally supposed, and the naive calculation differs from the answer provided by General Relativity by a factor of two. Nordström's theory, which

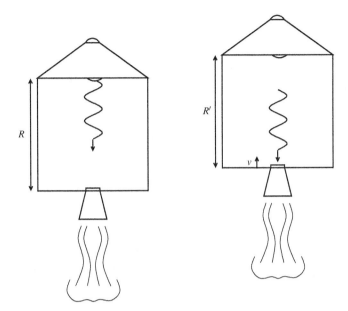

Figure 4.2   A spaceship of rest length R is accelerating upwards, while a light ray is sent backwards through it from the top, with frequency f in the inertial frame in which the ship is at rest at the instant the light ray is sent (left). By the time the ray arrives at the detector at the bottom (right), the ship is moving with speed v in the upward direction in the original frame. In a new inertial frame travelling with the ship at the instant of reception, the light has a Doppler-shifted frequency $f_{\text{rec}} = f(1 + v/c)$.

also obeys the Equivalence Principle, predicts no light bending at all.

However, there are two predictions for which the naive calculation is correct: that light should change wavelength in a gravitational field, and that clocks should run slower the deeper down a gravitational potential well they are. We go through these calculations next.

4.4.1   Gravitational redshift

Consider an accelerating spaceship, as shown in Figure 4.2. Light is emitted from the front towards the back. The frequency initially is f in an inertial frame travelling with the ship. The light takes a time $\Delta t = R/c$ to travel across the spaceship to be received at the back. But due to the acceleration, the back of the ship is now travelling at $v = a\Delta t$ in the original inertial frame. Hence, the light is received at the bottom of the ship with a Doppler-shifted frequency

$$f_{\text{rec}} = f\left(1 + \frac{\Delta v}{c}\right) = f\left(1 + \frac{a\Delta t}{c}\right) = f\left(1 + \frac{aR}{c^2}\right). \qquad (4.17)$$

By the Equivalence Principle, we must obtain the same result in a gravitational field producing gravitational acceleration g, so, as light moves downwards in a gravitational field, it should be blueshifted by a fraction $-gR/c^2$, where the minus sign indicates a higher frequency on reception. Now note that the change in gravitational potential between reception and emission is $\Delta\Phi = \Phi_{\text{rec}} - \Phi_{\text{emi}} = -gR$, so we can also write the relationship between the emitted and received frequencies as

$$f_{\text{rec}} = f\left(1 - \frac{\Delta\Phi}{c^2}\right). \qquad (4.18)$$

Conversely, light travelling out of the gravitational potential well of a body should be redshifted (see Problem 4.1) The largest effects are to be found in massive objects with small radii. Gravitational redshift was first conclusively observed in 1954 in the light from a white dwarf star; those have masses of about a Solar mass and radii of a few thousand kilometres.

4.4.2   Gravitational time dilation

An alternative view of the gravitational redshift is as a gravitational time dilation. This time let's direct a light ray upwards in a gravitational field, say from the surface of the Earth. Consider the passing of each wave crest as the tick of a clock with period $\Delta t(r) = 1/f(r)$. The light received by an observer at infinity is redshifted, and so the ticks there are spaced by longer time intervals $\Delta t(\infty)$. The relationship between the two follows directly from the redshift formula (4.18),

$$\Delta t(\infty) = \Delta t(r)\left(1 - \frac{\Phi(r)}{c^2}\right), \qquad (4.19)$$

where we have used the standard convention that the gravitational potential at infinity is zero, and so the nearer the gravitating object, the more negative the potential.

A distant observer looking at a clock within a gravitational field sees it running more slowly than the identical one they hold in their own hands. Conversely, an observer lower down in the gravitational potential sees distant clocks running faster. The effect is called gravitational time dilation. This is one of the General Relativistic effects that needs to be accounted for by the Global Positioning System so that it can ensure your navigation device is giving you the best route (see Problem 4.3).

4.5 Principles of the General Theory of Relativity

Having traced Einstein's thinking in developing a theory of gravity based on the Special Theory of Relativity and on the Equivalence Principle, along with some of the physical predictions such a theory can be expected to make, it is now time to condense it into a few principles which we will follow in developing the General Theory of Relativity.

> **Principle of General Relativity:** Physical laws are the same in all reference frames.

This extends the Principle of Special Relativity which stated that the laws of physics are to be the same in inertial reference frames. Now the demand is that the laws of physics are the same in all frames, not just inertial ones. Any observer is free to interpret physical laws using any coordinate system they choose, and all should agree on the outcome of any experiment. The necessity of extending the relativity principle was impressed on Einstein by his thinking about uniformly accelerating frames in the context of the Equivalence Principle, but this is a truly *general* principle, applying to all frames, not just ones which are in constant acceleration relative to an inertial frame.

> **Principle of General Covariance:** Equations expressing physical laws should take the same form under arbitrary coordinate transformations.

We will explore this in more detail in Chapter 5, but the essence was already there at the end of the previous one. The requirement of the invariance of the form of the physical laws is equivalent to demanding that the equations are expressed in terms of tensors, that is, quantities with well-defined behaviours under changes in coordinate systems, as tensors are defined by their form-invariance. This is essentially a mathematical formulation of the Principle of General Relativity.

> **The Correspondence Principle:** Old physical laws should be recovered in the appropriate limits.

When new physical laws are proposed, they may extend our knowledge into novel domains, but they will also often overlap with regimes where older laws have operated successfully. In those limiting cases, we would like to recover those older laws, now seen as approximations of our grander scheme. This is

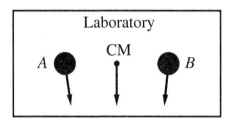

Figure 4.3   A laboratory in free-fall towards the Earth, containing two particles *A*, *B* that are initially at rest relative to the laboratory and its centre of mass CM. Over a short time interval, a measurement of the particle positions will not show a displacement. However, after a long enough time interval, or with a more accurate measurement, it will be seen that the particles move together. A small region around a freely falling trajectory should therefore behave like the space–time of Special Relativity, but show departures over larger distances and times.

what is meant by the Correspondence Principle. General Relativity, being both a theory of gravity and a guiding principle for construction of physical laws of all types, has a number of such limits. If we consider a situation without gravity, General Relativity should revert to Special Relativity. If the gravitational field is weak and objects are moving non-relativistically, Newton's longstanding laws of gravity ought to be recovered. In absence of both gravity and relativistic motions, Newtonian mechanics should apply.

We should be a little careful about what we mean by 'without gravity'. It means, for example, a region of space sufficiently far from any objects. But Einstein's Equivalence Principle makes clear that this also means any freely falling frame of reference. Thus astronauts in a space laboratory in free-fall around the Earth should be able to verify the laws of Special Relativity. However, if the laboratory is large enough, or the astronauts wait long enough, they will begin to see small deviations from Special Relativity.

For example, consider the laboratory falling towards the Earth in Figure 4.3. Two particles *A* and *B* placed at rest relative to the laboratory will fall with the same speed as the laboratory, and on a casual inspection, it will seem that they are obeying Special Relativity (and indeed Newtonian mechanics) by remaining at rest. However, a more careful measurement, either with higher accuracy in their relative positions or over a longer period of time, will reveal that they gradually move together, as all are on trajectories heading towards the centre of the Earth. If the particles are separated by s, and the laboratory is a distance R from the centre of the Earth, the tangential acceleration of the particles towards each other is

$$a_\mathrm{t} = s\frac{GM}{R^3}\,. \tag{4.20}$$

This can be compared with the radial acceleration $a_\mathrm{r} = GM/R^2$ experienced by both particles. The ratio goes to zero in proportion to s/R; particles very close together compared with R, the length scale over which the gravitational field changes, drift together very slowly.

Hence, we can expect that a sufficiently small region around a freely falling trajectory will behave like the Minkowski space–time of Special Relativity. However, we should not expect it to be possible to set up a coordinate system which behaves like Minkowski space–time everywhere. We will make 'sufficiently small' more precise after introducing the idea of curvature of space and space–time.

4.6 Towards curved space–time

We have learned that clocks run more slowly in a gravitational field. Inspecting equation (4.19), we can rewrite it as $\Delta t(r) = \Delta t(\infty)\left(1 + \Phi/c^2\right)$, provided the potential is small compared with c^2. The clock carried by an observer at radius r ticks more slowly than one carried by the observer at infinity. Recalling that the time recorded on an observer's clock is the proper time τ, we can write $d\tau = dt\left(1 + \Phi/c^2\right)$, where t is the time for the observer at infinity. The observer near the gravitating body is stationary at radius r; we can therefore write the suggestive equation:

$$d\tau^2 = -g_{\mu\nu}\,dx^\mu dx^\nu\,, \tag{4.21}$$

with

$$g_{00} = -\left(1 + \frac{2\Phi(r)}{c^2}\right)\,. \tag{4.22}$$

Hence, space–time in General Relativity should allow for more general metric tensors $g_{\mu\nu}$ than we encounter in Special Relativity. We have not yet specified what the other components of $g_{\mu\nu}$ might be, nor indeed, how they might be determined, but we do know that the Principle of General Covariance demands that we allow arbitrary coordinate transformations. This means that if we accept that one component of the metric tensor varies with position, we should expect that functions of the coordinates could appear in any of the components of the metric tensor. The space–time interval should therefore be written as

$$ds^2 = g_{\mu\nu}(x)\,dx^\mu dx^\nu\,, \tag{4.23}$$

to indicate that all components should be allowed to depend on the space–time position.

The consequences of allowing the metric to be a general function of space–time location are quite profound; in particular, we will find that space–time becomes an active participant in physical processes, and that it acquires fascinating geometrical properties including **curvature**. We will need some powerful mathematical tools to quantify space–time curvature, and to develop a physical law which explains how the curvature arises. But for now, we will begin to build some intuition about the metric and curvature using a simple example.

4.7   Curved space in two dimensions

It is hard to visualise four space–time dimensions, so we start by considering two-dimensional spaces, writing the distance between infinitesimally separated points $d\ell$ as

$$d\ell^2 = g_{ab}(x)dx^a dx^b\,, \tag{4.24}$$

where $a = 1, 2$ are indices labelling the two coordinates x^a.

First of all, it is important to distinguish between **intrinsic** and **extrinsic** curvature. Intrinsic curvature is a property of the surface itself, and can be determined without leaving the surface. Extrinsic curvature is a property of a surface embedded in a space of higher dimension. The General Theory of Relativity is formulated using intrinsic curvature only.

For example, a cylinder or torus appears to be curved, as we often think of it as a rolled-up sheet of paper, which is a three-dimensional point of view. It has extrinsic curvature, but no intrinsic curvature. If you live on such a surface, the properties of any shapes you draw will turn out to be the same as on a flat sheet of paper.[2] On a sphere, however, we would find for instance that the sum of angles of a triangle is more than $180°$, and that the circumference of a circle is less than $2\pi r$. These properties distinguish the sphere from a plane, without us

---

[2]More strictly, we are talking here of local geometric quantities. A cylinder is different from an infinite sheet in that you can go round the cylinder to return to your starting point, but that is a non-local property of the surface which relates to its **topology**, not its geometry. Topology in General Relativity is a fascinating topic, which might be relevant in several domains including cosmology and quantum gravity, but it is beyond the scope of this book.

ever needing to leave the surface. The latter is especially useful for defining the curvature at a point, known as the radius of curvature.

Let's consider some examples in two dimensions:

1. **The Euclidean plane.** In Cartesian coordinates x, y, the length element of the plane is given by $d\ell^2 = dx^2 + dy^2$. The metric is as simple as possible:

$$g_{ab} = \begin{pmatrix} 1 & 0 \\ 0 & 1 \end{pmatrix}. \tag{4.25}$$

In other coordinates, the simple properties of the plane are not so obvious. For example, in polar coordinates $x = r\cos\phi$, $y = r\sin\theta$ so that $dx = \cos\phi\, dr - r\sin\phi\, d\phi$ and $dy = \sin\phi\, dr + r\cos\phi\, d\phi$. The length interval is, therefore,

$$d\ell^2 = dr^2 + r^2 d\phi^2. \tag{4.26}$$

The metric is now position dependent, with

$$g_{ab} = \begin{pmatrix} 1 & 0 \\ 0 & r^2 \end{pmatrix}. \tag{4.27}$$

The change of coordinates doesn't alter the fact that we are still describing the same flat surface. We conclude that just because the metric is space dependent doesn't necessarily make the space curved.

2. **The sphere.** The length element on a sphere is given by

$$d\ell^2 = a^2 d\theta^2 + a^2 \sin^2\theta\, d\phi^2, \tag{4.28}$$

where a is constant giving the radius of the sphere, so that

$$g_{ab} = \begin{pmatrix} a^2 & 0 \\ 0 & a^2\sin^2\theta \end{pmatrix}. \tag{4.29}$$

This space is genuinely curved; even though a small region of the sphere is approximately Euclidean, it is impossible to find coordinates which make the metric Euclidean everywhere. To see why, let's look at a way telling whether a space is curved or not, beginning our exploring of curvature.

In order to illustrate the mathematical idea of curvature, let us continue with the example of the sphere. Suppose we draw a small circle around the North Pole, with coordinates $\theta = \theta_0$ and $0 \leq \phi < 2\pi$ (see Figure 4.4). The metric tells us that

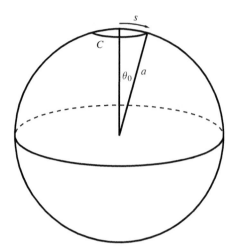

Figure 4.4   The circumference C of a circle of radius s on a sphere of radius a, as given in equation (4.30), is less than $2\pi s$, From this, we can derive a measure of the curvature of the surface at any point.

the radius as measured on the surface of the sphere is $s = \int d\ell = a \int_0^{\theta_0} d\theta = a\theta_0$, and its circumference is

$$C(s) = a\sin\theta_0 \int_0^{2\pi} d\phi = 2\pi a \sin\frac{s}{a} = 2\pi s \left(1 - \frac{s^2}{6a^2} + \cdots \right). \qquad (4.30)$$

The circumference of a circle on the sphere is therefore less than the Euclidean value $2\pi s$ for any finite s, but $C \to 2\pi s$ in the limit $s \to 0$.

From this limit, we can infer that the smaller the region of the sphere is, the more like a plane it is. Indeed, in terms of a radial coordinate $\rho = a\sin\theta$, the length element is given approximately by

$$d\ell^2 \simeq d\rho^2 + \rho^2 d\phi^2, \qquad (4.31)$$

for $s \ll a$. This is just the length element of the plane in polar coordinates. For that reason, in your daily life, you are not likely to notice that the Earth's surface is globally curved, unless you are taking a long plane flight.

However, even though the difference disappears in the limit $s \to 0$, there is still information in $C(s)$ about the radius of the sphere a. This information can be extracted by taking derivatives of the circumference as a function of radius. Let us define a quantity

$$K = -\lim_{s \to 0} \frac{C''(s)}{C(s)} = \frac{1}{a^2}. \tag{4.32}$$

This quantity, called the **Gaussian curvature**, is a measure of the curvature of space. It is defined at a particular point, the location where we centred our circle; thus, it is a local quantity. It is an *intrinsic* curvature, which means that it depends entirely on quantities evaluated within the surface itself, and their derivatives. The existence of an intrinsic measure of the curvature was demonstrated by Carl Friedrich Gauss in his 'Remarkable Theorem' (*Theorema Egregium*). As it is a local quantity, on a general two-dimensional surface, it could vary from point to point.

The intrinsic curvature of four-dimensional space–time is one of the fundamental objects in General Relativity. In Chapter 5, we will assemble the machinery to define and calculate it for a general space–time metric $g_{\mu\nu}(x)$. The two important points to take away from this section are that

- A sufficiently small region of a space is approximately Euclidean. This is the two-dimensional analogue of the statement that in a small-enough region of a space–time, the metric is well approximated by the Minkowski metric of Special Relativity.

- There is a geometrical quantity called intrinsic curvature, which is local and is computable from the metric and its derivatives.

Problems

4.1* The frequency f of a photon climbing out of a weak gravitational field is given by

$$f = f_{em} \left(1 - \frac{\Delta\Phi}{c^2} \right),$$

where f_{em} is the frequency at emission and $\Delta\Phi$ is the change in gravitational potential between the emission and detection points:

a) If the photon is emitted at the surface of a spherical body of radius r, show that the frequency measured by an observer at infinity is

$$f(\infty) = f_{em} \left(1 - \frac{GM}{rc^2} \right).$$

b) Calculate the relative frequency change $\Delta f/f_{em}$ for the white dwarf star

Sirius B, which can be taken to have mass 1.0 $M_\odot$ and radius 5800 km. Express the result as the velocity which would give the same Doppler shift.

c) What would be the frequency of light observed at infinity if the radius of the body were $r = GM/c^2$? If M is the mass of the Earth, what is the value of this radius? Why should we be suspicious of this result?

4.2   It is tempting to apply the Equivalence Principle to zero-mass particles, in which case photons should also fall in a gravitational field, causing a bending of the light ray. Consider a ray travelling horizontally at the Earth's surface, and take the acceleration due to gravity as $10 \mathrm{m\,s}^{-2}$ and $c = 3 \times 10^8 \mathrm{ms}^{-1}$.

a) After it has travelled 10 km, how far has the light fallen vertically?

b) Over the same distance, how far has the Earth's surface curved away due to the Earth being a sphere (approximate the Earth as a perfect sphere of radius 6500 km)?

c) By what factor would the mass of the Earth have to be increased to make the answers to (a) and (b) the same, and what would this mean?

d) The above calculation gives a deflection of light differing by a factor of 2 from General Relativity. Suggest a reason to doubt the calculation.

4.3   Global Positioning System satellites orbit the Earth at a radial distance from the centre of about 30 000 km.

a) Estimate the rate at which the clock on a GPS satellite runs faster than one on the Earth due to the gravitational redshift, expressing the result in seconds per day. What distance error would this cause after one hour?

b) There is also a time dilation effect due to the satellite's speed relative to a non-rotating frame tied to the centre of the Earth. Assuming the satellite is in a circular orbit, calculate the difference in the rates at which the clocks run in the two frames.

4.4   Explain why there can be no bending of light around massive objects in Nordström's theory of gravitation.

4.5   Consider a two-dimensional space with a length element

$$d\ell^2 = R^2 d\eta^2 + R^2 \sinh^2 \eta \, d\phi^2,$$

where R is constant. Write down the metric, and calculate the intrinsic curvature at the point $\eta = 0$.

4.6* Consider the length element in three-dimensional Euclidean space,

$$d\ell_3^2 = dx^2 + dy^2 + dz^2 \,.$$

a) Find the length element in spherical polar coordinates r, θ, ϕ, with $x = r\sin\theta\cos\phi$, $y = r\sin\theta\sin\phi$, and $z = r\cos\theta$. Hence, show that equation (4.28) is the length element for the spherical surfaces $r = a$.

b) Defining the coordinate $\rho = \sqrt{x^2 + y^2}$, show that the length element on the sphere at $r = a$ is

$$d\ell^2 = \left(1 - \frac{\rho^2}{a^2}\right)^{-1} d\rho^2 + \rho^2 d\phi^2 \,.$$

c) Calculate the Laplacian operator ∇^2 for functions on the sphere in these new coordinates, $f(\rho,\phi)$. Check that it becomes the ordinary two-dimensional Laplacian operator in polar coordinates in the limit $\rho \ll a$, i.e.

$$\nabla^2 \to \frac{1}{\rho}\frac{\partial}{\partial\rho}\left(\rho\frac{\partial}{\partial\rho}\right) + \frac{1}{\rho^2}\frac{\partial^2}{\partial\phi^2} \,.$$

Chapter 5

Tensors and Curved Space–Time

*Use any coordinates you want · tensors ensure the laws remain the
same · differentiation becomes trickier ... · ... but then lets us define
curvature*

5.1 General coordinate transformations

The Lorentz transformation was a linear transformation between coordinate systems. In General Relativity, we have to consider transformations in which the new coordinates are arbitrary invertible functions of the old ones:

$$x^{\mu} \to x'^{\mu}(x^0, x^1, x^2, x^3). \tag{5.1}$$

A useful example to have in mind is a transformation between Cartesian and polar coordinates, for example in two dimensions $(r, \theta) \to (x, y)$, where $x = r\cos\theta$ and $y = r\sin\theta$. It is crucial to understand that a change of coordinates does not change the space–time itself, only the way in which we label it. Moreover, while all coordinate systems are in principle equally valid, actual calculations may be much simpler in a well-chosen coordinate system that reflects the symmetry of the situation under study. It might also be the case that some coordinate systems are capable of covering only a portion of the space–time; we will see an example in the black-hole space–time of Chapter 11.

Fortunately, the general formalism is actually intuitively rather simpler than the Lorentz transformation, because it simply takes the form of the chain rule for

Introducing General Relativity, First Edition. Mark Hindmarsh and Andrew Liddle.
© 2022 John Wiley & Sons Ltd. Published 2022 by John Wiley & Sons Ltd.
Companion website: www.wiley.com/go/hindmarsh/introducingGR

partial derivatives. The transformation law of coordinate differentials is

$$dx^\mu \to dx'^\mu = \left(\frac{\partial x'^\mu}{\partial x^\nu}\right) dx^\nu , \tag{5.2}$$

where as always there is an implied sum in the repeated indices.[1]

The Lorentz transformation $dx'^\mu = \Lambda^\mu{}_\nu dx^\nu$ is a special case; since it is a linear transformation, the partial derivatives in that case are constants independent of the coordinates.

Similarly, derivatives transform as

$$\partial_\mu \equiv \frac{\partial}{\partial x^\mu} \to \frac{\partial}{\partial x'^\mu} = \frac{\partial x^\nu}{\partial x'^\mu} \frac{\partial}{\partial x^\nu} . \tag{5.3}$$

Note that a raised index on the denominator is equivalent to a lowered index on the numerator.

Contravariant 4-vectors are any vectors whose components transform like dx^μ, and covariant 4-vectors are vectors whose components transform like $\partial_\mu \equiv \partial/\partial x^\mu$. Hence, the laws are

$$a^\mu \to a'^\mu = \frac{\partial x'^\mu}{\partial x^\nu} a^\nu ; \tag{5.4}$$

$$b_\mu \to b'_\mu = \frac{\partial x^\nu}{\partial x'^\mu} b_\nu . \tag{5.5}$$

Be careful to note that the transformation terms are still partial derivatives of the coordinates. These are straightforward generalisations of the relations we saw earlier.

For tensors of a higher rank, there is again one transformation term for each of the indices, taking the appropriate form depending on whether the index is contravariant or covariant. The metric tensor $g_{\mu\nu}$ is an example of a rank-two covariant tensor. It satisfies

$$g_{\mu\nu} \to g'_{\mu\nu} = \frac{\partial x^\rho}{\partial x'^\mu} \frac{\partial x^\sigma}{\partial x'^\nu} g_{\rho\sigma} . \tag{5.6}$$

This relation guarantees the invariance of the interval ds^2, as the transformation terms from the metric will cancel out the transformation terms from the coordinates. The interval is a scalar quantity.

The metric tensor is used to lower indices, and we raise indices with its inverse $g^{\mu\nu}$ defined by $g^{\mu\nu} g_{\nu\rho} = \delta^\mu_\rho$. Unlike in Special Relativity, the inverse metric

---

[1] See Problem 4.6 for some practice with coordinate transformations.

typically does not have the same components as the metric. For the particular case of a metric whose only non-zero terms are on the diagonal (i.e. when each index has the same value), the components of $g^{\mu\nu}$ are the reciprocals of those of $g_{\mu\nu}$.

5.2 Tensor equations and the laws of physics

The Principle of General Covariance stated that the laws of physics should be tensor equations. We are now in a better position to see why. An equation relating tensors will have a certain rank, corresponding to the unsummed indices. Group all the terms on the left-hand side of the equation. If we change coordinate system, the fact that the indices must match means that each term picks up the same set of transformation terms, of the form $(\partial x'^{\mu}/\partial x^{\nu})$ or $(\partial x^{\nu}/\partial x'^{\mu})$ depending on whether indices are raised or lowered. The transformed left-hand side is therefore the original one with an overall multiplier of the transformation terms. Hence, the sum of terms equalling zero in the original coordinate system clearly implies that the same sum in the new coordinate system also does.

If we write all our equations as tensor equations, we are therefore guaranteed that they take the same form in any coordinate system, i.e. that the laws of physics are the same for all observers.

The rules for forming tensor equations are as follows:

1. The product of two tensors is also a tensor. For example, if $F^{\mu\nu}$ and u_{ρ} are both tensors, then so is $F^{\mu\nu}u_{\rho}$. More precisely, it is a tensor of rank $\binom{2}{1}$.

2. The contraction of two indices on a tensor is also a tensor; following the previous example, $F^{\mu\nu}u_{\nu}$ is a tensor with rank $\binom{1}{0}$.

3. Sums and difference of tensors of the same rank are also tensors. Hence, all terms in a tensor equation must be tensors of the same rank.

4. If the contraction of an index of an object $U^{\mu\cdots\nu}{}_{\rho\cdots\sigma}$ with a tensor is also a tensor, then $U^{\mu\cdots\nu}{}_{\rho\cdots\sigma}$ is itself a tensor. This is known as the quotient law.

5.3 Partial differentiation of tensors

Differentiation is an essential operation when formulating laws of physics, and we need a covariant way of doing this if we are to construct physical laws consistent with the Principle of General Covariance.

In General Relativity, physical laws are formulated in terms of tensor fields, that is, tensors which are functions of the space and time. Let us start our study of differentiation of tensor fields by considering the derivative of a contravariant 4-vector field a^μ. To define a derivative, we need to consider the difference in its value between two neighbouring points. Normally, the space and time derivatives of a vector field would be defined as

$$\frac{\partial a^\mu}{\partial x^\nu} \equiv \lim_{\delta x^\nu \to 0} \frac{a^\mu(x^\nu + \delta x^\nu) - a^\mu(x^\nu)}{\delta x^\nu}. \tag{5.7}$$

But is this a tensor, as it needs to be for use in physical laws?

We can check whether it is by examining the differential da^μ in the new coordinates, as the chain rule gives

$$da^\mu = \frac{\partial a^\mu}{\partial x^\sigma} dx^\sigma. \tag{5.8}$$

The differential in the new coordinates is

$$\begin{aligned} da'^\mu &= d\left(\frac{\partial x'^\mu}{\partial x^\nu} a^\nu\right) \\ &= \frac{\partial x'^\mu}{\partial x^\nu} da^\nu + \frac{\partial^2 x'^\mu}{\partial x^\nu \partial x^\sigma} a^\sigma dx^\nu. \end{aligned} \tag{5.9}$$

The first piece is what is required for da^μ to be a 4-vector, but the second piece spoils it. Hence, if da^μ is not a 4-vector, then neither is $\partial a^\mu / \partial x^\nu$, and so it is not suitable for the formulation of physical laws. Note that this wouldn't be a problem in Special Relativity; the linearity of the Lorentz transformation means that the second term would vanish as the second derivative is zero.

5.4   The covariant derivative and parallel transport

We saw that the simple differential of a vector da^μ does not transform linearly. We aim to *define* a differential Da^μ which does transform as a vector, and generalise this to all tensors. From this covariant differential, we can define the covariant derivative of a tensor, which is itself a tensor, but of higher rank.

In essence, we need a new term whose transformation will cancel out the extra term we found above. To this end, we define

$$Da^\mu \equiv da^\mu + \Gamma^\mu_{\nu\sigma} a^\nu dx^\sigma, \tag{5.10}$$

where the new object $\Gamma^\mu_{\nu\sigma}$ is known as the **connection** and is constructed to have the correct transformation property to precisely cancel the second term in equation (5.9) which spoiled the tensor nature of da^μ.

It is a fundamental theorem of Riemannian geometry that the connection can be chosen so that its components are symmetric in its last two indices,

$$\Gamma^{\mu}_{\nu\sigma} = \Gamma^{\mu}_{\sigma\nu}. \tag{5.11}$$

Furthermore, the connection can also be chosen so that the covariant derivative of the metric tensor vanishes, a condition called **metric compatibility**. Making these two choices gives a connection known as the **Levi-Civita connection**, and its components $\Gamma^{\mu}_{\nu\sigma}$ the **Christoffel symbols**. The standard formulation of General Relativity, which we are following, is founded on choosing the symmetric Levi-Civita connection.

We will shortly see how to construct the covariant derivative of the metric, and also why it is important that it vanishes. The choices of metric compatibility and symmetry uniquely fix the form of the Christoffel symbols to be (see Problem 5.7)

$$\Gamma^{\mu}_{\nu\sigma} = \frac{1}{2}g^{\mu\rho}\left(\partial_{\sigma}g_{\rho\nu} + \partial_{\nu}g_{\rho\sigma} - \partial_{\rho}g_{\nu\sigma}\right). \tag{5.12}$$

One can check, in a lengthy calculation, that this quantity does indeed transform in the correct way to cancel the second term in equation (5.9).

We make the following observations:

- The Christoffel symbols are not the components of a tensor, though they can be used to construct tensors. Indeed, they can't be tensor components, otherwise, they would not be able to compensate for the failure of da^{μ} to be one.

- As they are not tensor components you cannot shift the indices up and down; it should always appear with one raised and two lowered.

- In Minkowski space–time, $\Gamma^{\mu}_{\nu\sigma} = 0$ in Cartesian coordinates, and the covariant derivative becomes the ordinary derivative. This can be considered an example of the Correspondence Principle.

- The components $\Gamma^{\mu}_{\nu\sigma}$ may not vanish even in Minkowski space–time, if the coordinates are curvilinear (e.g. spherical polar coordinates).

- Strictly speaking, we are discussing the Christoffel symbols of the second kind. There are also rarely used Christoffel symbols of the first kind, $[\rho\nu,\sigma] = \frac{1}{2}(\partial_{\sigma}g_{\rho\nu} + \partial_{\nu}g_{\rho\sigma} - \partial_{\rho}g_{\nu\sigma})$.

There is an important geometric interpretation to the connection. In order to define the differential of a vector field, we need to take the difference between vectors at infinitesimally separated points x^{μ} and $x^{\mu} + dx^{\mu}$. However, to do this,

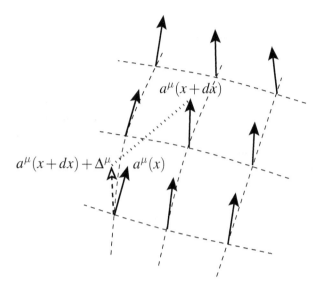

Figure 5.1 In order to define covariant differentiation of a vector field $a^\mu(x)$ in a space–time with a curvilinear coordinate system, represented by the grid of dashed lines, the vector at space–time point $x+dx$ must be moved to a point x in a way which preserves its magnitude and direction. This is called parallel transport. The vector field at $x+dx$ parallel transported to x is shown as a dashed arrow; its components $a^\mu(x+dx)$ receive a small correction Δ^μ.

we need to bring the vector at $x^\mu + dx^\mu$ to the point x^μ, while preserving its magnitude and direction. This process is called **parallel transport**. In a general coordinate system, the components of a vector field change under parallel transport by a small amount Δ^μ (see Figure 5.1). This change is proportional to both the value of the vector $a^\mu(x+dx)$ and to the displacement dx^μ. The proportionality coefficients are the components of the connection.

Hence, under parallel transport from $x+dx$ to x, the components of the vector $a(x+dx)$ change by

$$a^\mu(x+dx) \to a^\mu(x+dx)+\Delta^\mu(a,dx) = a^\mu(x+dx)+\Gamma^\mu_{\nu\sigma}a^\nu(x+dx)dx^\sigma. \quad (5.13)$$

We can neglect the difference between $a(x)$ and $a(x+dx)$ in the last term as it produces terms of higher order in dx. The covariant differential of a vector field is therefore

$$Da^\mu(x) = a^\mu(x+dx)+\Delta^\mu(a,dx)-a^\mu(x) = da^\mu +\Gamma^\mu_{\nu\sigma}a^\nu(x)dx^\sigma. \quad (5.14)$$

Note that if $Da^\mu(x) = 0$, then the vectors $a^\mu(x+dx)$ and $a^\mu(x)$ are by definition related by parallel transport.

We then define the **covariant derivative** as the tensor

$$a^\mu{}_{;\sigma} \equiv \frac{Da^\mu}{dx^\sigma}. \tag{5.15}$$

Note the use of the semicolon to indicate the coordinate in which the derivative is taken. In order to keep expressions compact, we will also use a comma to indicate a partial derivative, e.g. $\phi_{,\mu} \equiv \partial_\mu \phi$. Hence,

$$a^\mu{}_{;\sigma} = a^\mu{}_{,\sigma} + \Gamma^\mu{}_{\nu\sigma} a^\nu. \tag{5.16}$$

An alternative notation for the covariant derivative is

$$\nabla_\sigma a^\mu \equiv a^\mu{}_{;\sigma}, \tag{5.17}$$

which mirrors the use of ∂_σ for the partial derivative. We shall mostly use the semicolon notation.

There is no difficulty in comparing the values of a scalar at neighbouring points; the partial derivative of a scalar already has the correct transformation property to be a vector. The partial derivative of a scalar is therefore already a covariant derivative, so as a matter of notation

$$\phi_{;\rho} = \phi_{,\rho}. \tag{5.18}$$

One can use this property on the scalar $\phi = a^\mu b_\mu$ to show that the components of the covariant derivative of a covariant vector b_μ are

$$b_{\mu;\sigma} = b_{\mu,\sigma} - \Gamma^\nu{}_{\sigma\mu} b_\nu. \tag{5.19}$$

Problem 5.9 asks you to do the detailed working to demonstrate this.

To discover the rule for covariantly differentiating a higher-rank tensor, we can consider the special case of tensors which are the product of two vectors. The rule for a general higher-rank tensor cannot be any different. Suppose we construct a rank-two tensor $M^{\mu\nu}$ from two contravariant vectors,

$$M^{\mu\nu} = a^\mu b^\nu. \tag{5.20}$$

Let us examine the differential of the tensor, as we did for the vector a^μ:

$$dM^{\mu\nu} = (da^\mu)b^\nu + a^\mu(db^\nu). \tag{5.21}$$

In a different coordinate system x', each of the differentials da^μ and db^ν will fail to be a tensor in the way we discovered around equation (5.9). We can anticipate that in order to make a covariant object, we will need to use the Christoffel symbol. Let us write the differentials as the difference between the covariant differential and the parallel transport term,

$$dM^{\mu\nu} = (Da^\mu - \Gamma^\mu{}_{\rho\sigma} a^\rho dx^\sigma) b^\nu + a^\mu (Db^\nu - \Gamma^\nu{}_{\rho\sigma} b^\rho dx^\sigma). \qquad (5.22)$$

We can rearrange this equation to give an expression which is clearly covariant on the right-hand side

$$dM^{\mu\nu} + \Gamma^\mu{}_{\rho\sigma} a^\rho dx^\sigma b_\nu + a^\mu \Gamma^\nu{}_{\rho\sigma} b^\rho dx^\sigma = (Da^\mu) b^\nu + a^\mu (Db^\nu). \qquad (5.23)$$

The left-hand side must also be covariant and defines the covariant differential of this rank-$\binom{2}{0}$ tensor

$$DM^{\mu\nu} = dM^{\mu\nu} + \Gamma^\mu{}_{\rho\sigma} M^{\rho\nu} dx^\sigma + \Gamma^\nu{}_{\rho\sigma} M^{\mu\rho} dx^\sigma. \qquad (5.24)$$

Therefore, the components of the covariant derivative of the tensor are

$$M^{\mu\nu}{}_{;\sigma} = M^{\mu\nu}{}_{,\sigma} + \Gamma^\mu{}_{\rho\sigma} M^{\rho\nu} + \Gamma^\nu{}_{\rho\sigma} M^{\mu\rho}. \qquad (5.25)$$

As might have been expected, the differentiated tensor acquires one Christoffel symbol per index.

The rule for covariant differentiation of mixed or covariant tensors can be derived, for example, by covariantly differentiating the vectors $c^\mu = M^\mu{}_\nu b^\nu$ and $d_\mu = M_{\mu\nu} b^\nu$, and rearranging to isolate manifestly covariant expressions $M^\mu{}_\nu b^\nu{}_{;\rho}$ and $d_\mu = M_{\mu\nu} b^\nu{}_{;\rho}$ on one side of the equations. The results are

$$M^\mu{}_{\nu;\sigma} = M^\mu{}_{\nu,\sigma} + \Gamma^\mu{}_{\rho\sigma} M^\rho{}_\nu - \Gamma^\rho{}_{\nu\sigma} M^\mu{}_\rho; \qquad (5.26)$$

$$M_{\mu\nu;\sigma} = M_{\mu\nu,\sigma} - \Gamma^\rho{}_{\mu\sigma} M_{\rho\nu} - \Gamma^\rho{}_{\nu\sigma} M_{\mu\rho}. \qquad (5.27)$$

We can infer that the general rule for the covariant derivative of tensors is that each index on a tensor picks up a set of Christoffel symbols with the sign positive or negative, and the contracted index down or up, depending on whether the index is contra- or covariant.

Finally, we summarise a few important properties of the covariant derivative. Proofs are left as exercises in the problems at the end of the chapter.

- The covariant derivative of the metric vanishes (the metric-compatibility condition)

$$g_{\mu\nu;\rho} = 0.$$ (5.28)

- Covariant differentiation commutes with raising and lowering of indices (a corollary of the previous property, showing why metric compatibility is so important).

$$a_{\mu;\sigma} = g_{\mu\nu} \left(a^{\nu}{}_{;\sigma} \right).$$ (5.29)

- The index on a covariant derivative can be raised:

$$a_{\mu}{}^{;\sigma} = g^{\sigma\nu} a_{\mu;\nu}.$$ (5.30)

- Covariant derivatives obey a product rule (see Problem 5.5).

$$\left(a^{\mu} b^{\nu} \right)_{;\rho} = a^{\mu}{}_{;\rho} b^{\nu} + a^{\mu} b^{\nu}{}_{;\rho}.$$ (5.31)

- Covariant derivatives do not generally commute unless space–time is flat: for example typically $a^{\mu}{}_{;\alpha;\beta} \neq a^{\mu}{}_{\beta;\alpha}$. An exception is two covariant derivatives applied to a scalar: $\phi_{,\alpha;\beta} = \phi_{\beta;\alpha}$, as the first derivative is equivalent to an ordinary derivative (see Problem 5.6).

5.5 Christoffel symbols of a two-sphere

As an example of evaluating Christoffel symbols, we will study a familiar curved surface, the two-dimensional sphere, or two-sphere. The line element $d\ell$ on a two-sphere of radius a is given in polar coordinates by

$$d\ell^2 = a^2 d\theta^2 + a^2 \sin^2\theta \, d\phi^2.$$ (5.32)

The non-zero metric components are therefore the diagonal entries $g_{\theta\theta} = a^2$ and $g_{\phi\phi} = a^2 \sin^2\theta$. For diagonal metrics, the components of the inverse are also diagonal, and are just the inverses of the corresponding components: $g^{\theta\theta} = 1/a^2$ and $g^{\theta\theta} = 1/a^2 \sin^2\theta$.

From the formula for the Christoffel symbols, and using the indices a, b, etc. to label the two coordinates, we have

$$\Gamma^a{}_{bc} = \frac{1}{2} g^{ad} \left(\partial_c g_{db} + \partial_b g_{dc} - \partial_d g_{bc} \right).$$ (5.33)

We can now work out all the components. Each index can take one of two values, θ and ϕ, so there are eight permutations in all. They are

$$
\begin{aligned}
\Gamma^{\theta}{}_{\theta\theta} &= \frac{1}{2a^2}\partial_{\theta}g_{\theta\theta} = 0; \\[2mm]
\Gamma^{\phi}{}_{\phi\phi} &= \frac{1}{2a^2}\partial_{\phi}g_{\phi\phi} = 0; \\[2mm]
\Gamma^{\theta}{}_{\phi\phi} &= -\frac{1}{2a^2}\partial_{\theta}g_{\phi\phi} = -\sin\theta\,\cos\theta\,; \\[2mm]
\Gamma^{\phi}{}_{\theta\theta} &= 0; \\[2mm]
\Gamma^{\phi}{}_{\theta\phi} &= \Gamma^{\phi}{}_{\phi\theta} = \frac{\cos\theta}{\sin\theta}\,; \\[2mm]
\Gamma^{\theta}{}_{\theta\phi} &= \Gamma^{\theta}{}_{\phi\theta} = 0.
\end{aligned}
\tag{5.34}
$$

5.6   Parallel transport on a two-sphere

We can use our results from Section 5.5 to show how a vector v^a behaves under parallel transport on the sphere. Recalling the covariant differential, the equation for parallel transport in a direction dx^c is

$$
\frac{Dv^a}{dx^c} = \partial_c v^a + \Gamma^a{}_{bc}v^b = 0.
\tag{5.35}
$$

Let us start with the easiest case of parallel transport in the direction of the polar angle, $d\theta$. We can then write

$$
\frac{\partial v^{\phi}}{\partial\theta} = 0,
\tag{5.36}
$$

$$
\frac{\partial v^{\theta}}{\partial\theta} + \Gamma^{\phi}{}_{\phi\theta}v^{\phi} = 0.
\tag{5.37}
$$

Let us suppose we start at the equator with components v^{θ}_e, v^{ϕ}_e. Equation (5.36) tells us that the θ-component of the vector remains constant, while the solution to equation (5.37) shows that the ϕ-component behaves as follows:

$$
v^{\phi}(\theta) = \frac{v^{\phi}_e}{\sin\theta}
\tag{5.38}
$$

with the polar angle θ. So, for example, a vector pointing due east at the equator ($v^{\theta}_e = 0$) will stay pointing east when parallel transported towards a pole (see Figure 5.2).

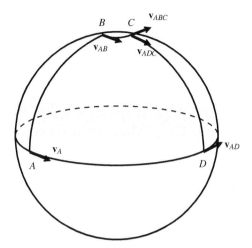

Figure 5.2 Parallel transport of a vector on a two-sphere. A vector $\mathbf{v}_A$ pointing east at the equator will stay pointing east as it is parallel transported towards a pole along a line of longitude AB. When parallel transported along an arc of a small circle BC near a pole, the vector $\mathbf{v}_{AB}$ rotates to point south ($\mathbf{v}_{ABC}$). Parallel transport along the equator AD does not change the direction of the vector, so $\mathbf{v}_{AD}$ still points east, as does $\mathbf{v}_{ADC}$, the vector after parallel transport along a line of longitude to C. The difference in directions of the vectors $\mathbf{v}_{ABC}$ and $\mathbf{v}_{ADC}$, which have been parallel transported from A to C along different paths, indicates that the sphere has curvature.

Note that the ϕ-component diverges at the poles ($\theta \to 0, \pi$), but the magnitude of the vector remains constant, as

$$\mathbf{v}^2 = \left(v^\theta\right)^2 g_{\theta\theta} + \left(v^\phi\right)^2 g_{\phi\phi} = a^2 \left[\left(v_{\mathrm{e}}^\theta\right)^2 + \left(v_{\mathrm{e}}^\phi\right)^2\right]. \tag{5.39}$$

This illustrates one of the properties of parallel transport: the preservation of magnitude of the vector.

We can also study the components along the unit vectors given by $\mathbf{e}_\theta = (1,0)/a$ and $\mathbf{e}_\phi = (0,1)/a\sin\theta$,

$$\mathbf{e}_\theta \cdot \mathbf{v} = a v_{\mathrm{e}}^\theta, \quad \mathbf{e}_\phi \cdot \mathbf{v} = a v_{\mathrm{e}}^\phi, \tag{5.40}$$

which remain constant in this case of parallel transport along a line of longitude.

Parallel transport along a line of latitude is more complicated (see Problem 5.4), but a similar procedure shows that a vector pointing due east at $\theta = \theta_0$ and

$\phi = 0$, so having components $v^\theta = 0$ and $v^\phi = v_0^\phi$, changes with latitude as

$$v^\theta(\phi) = v_0^\phi \sin\theta_0 \sin[(\cos\theta_0)\phi]; \tag{5.41}$$
$$v^\phi(\phi) = v_0^\phi \cos[(\cos\theta_0)\phi]. \tag{5.42}$$

Hence, an east-pointing vector parallel transported in a small circle around the north pole makes a 2π rotation in the limit $\theta_0 \to 0$. In Figure 5.2, this rotation is illustrated by the path BC, where an east-pointing vector $\mathbf{v}_{AB}$ comes to point south ($\mathbf{v}_{ABC}$) after an approximately $\pi/2$ change in longitude near a pole. An east-pointing vector does not change direction when parallel transported along the equator (path AD).

Note that vectors parallel transported along two different paths between two points come to point in different directions, as illustrated by the vectors $\mathbf{v}_{ABC}$ and $\mathbf{v}_{ADC}$ at point C, which have taken the paths ABC and ADC, respectively. This important feature will help us with the formal definition of curvature in Section 5.7.

5.7   Curvature and the Riemann tensor

In Section 5.6, we saw that when a vector is parallel transported between two points around two different paths on a curved space, it arrives pointing in different directions according to the path taken.

This behaviour can be used to define curvature in space–time as well as on the two-sphere. Suppose we take a vector $a(x)$ from point A with coordinates x to point B with coordinates $x + dx_1$ and then to a further point C with coordinates $x + dx_1 + dx_2$, as shown in Figure 5.3. This results in changes to the vector, which can be written as

$$a(x) \to a(x) + D_1 a(x) \to a(x) + D_1 a(x) + D_2(a(x) + D_1 a(x)). \tag{5.43}$$

If we instead parallel transport the vector via point D, so that the parallel transport operations are carried out in the opposite order, we obtain

$$a(x) \to a(x) + D_2 a(x) \to a(x) + D_2 a(x) + D_1(a(x) + D_2 a(x)). \tag{5.44}$$

The two vectors at the point $x + dx_1 + dx_2$ are different, because $D_1 D_2 a(x)$ is not equal to $D_2 D_1 a(x)$. The difference defines the **Riemann tensor** (also called the curvature tensor), through

$$D_1 D_2 a^\mu(x) - D_2 D_1 a^\mu(x) = R^\mu{}_{\nu\alpha\beta} a^\nu dx_1^\alpha dx_2^\beta. \tag{5.45}$$

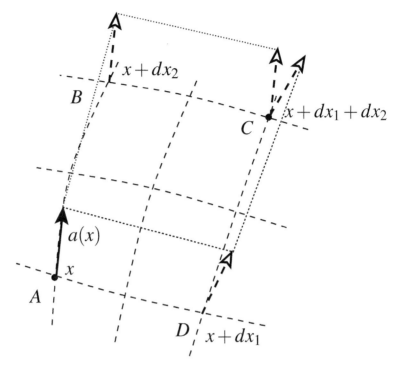

Figure 5.3 When a vector is parallel transported from point A to point C in two different ways, via B or D, in a curved space–time, the end result depends on the path taken. The difference in the two transported vectors defines the curvature tensor of the space–time.

From the definition of the covariant differential, and the arbitrariness of the infinitesimal coordinate differences dx_1^α and dx_2^β, it follows that

$$a^\mu{}_{;\alpha;\beta} - a^\mu{}_{;\beta;\alpha} = -R^\mu{}_{\nu\alpha\beta}a^\nu . \tag{5.46}$$

A fair amount of algebra leads to the Riemann tensor in terms of the Christoffel symbols as[2]

$$R^\mu{}_{\nu\alpha\beta} = \Gamma^\mu{}_{\nu\beta,\alpha} - \Gamma^\mu{}_{\nu\alpha,\beta} + \Gamma^\mu{}_{\rho\alpha}\Gamma^\rho{}_{\nu\beta} - \Gamma^\mu{}_{\rho\beta}\Gamma^\rho{}_{\nu\alpha} . \tag{5.47}$$

---

[2]Different books have different sign conventions, for instance some define the Riemann tensor in equation (5.45) without the minus sign, resulting in the opposite sign in the formula (5.47). You need to be extremely careful if you attempt to combine equations from different textbooks. For reference, our sign conventions match those of the books by Misner, Thorne, and Wheeler and by Schutz.

This algebra is the subject of Problem 5.10.

A similar equation applies to covariant vectors (see Problem 5.12)

$$a_{\mu;\alpha;\beta} - a_{\mu;\beta;\alpha} = R^{\nu}{}_{\mu\alpha\beta} a_{\nu}. \tag{5.48}$$

The Riemann tensor contains a complete description of the curvature of space–time. Its tensor nature follows from its definition given above in equation (5.46); this is a properly composed tensor equation from which the appropriate tensor transformation law for the Riemann tensor follows. It is harder to establish its transformation properties from the expression in terms of the Christoffel symbols (5.47).

Another way of thinking about the Riemann tensor and curvature comes from looking at the difference between the vector transported along the two different paths, $\delta a^{\mu} = R^{\mu}{}_{\nu\alpha\beta} a^{\nu} dx_1^{\alpha} dx_2^{\beta}$. It can be shown that $\delta a^{\mu} = R^{\mu}{}_{\nu\alpha\beta} a^{\nu} dx_1^{\alpha} dx_2^{\beta}$ is the same change one would get if one made an infinitesimal Lorentz transformation of the original vector. This is the four-dimensional space–time analogy with the relative rotation of vectors parallel transported along different paths on a two-dimensional sphere, investigated in Section 5.6.

In Minkowski space–time, the Riemann tensor vanishes everywhere, meaning that $R^{\mu}{}_{\nu\alpha\beta}(x) = 0$ for all space–time coordinates x. This is straightforward to see: the Minkowski metric in Cartesian coordinates $\eta_{\mu\nu}$ gives vanishing Christoffel symbols $\Gamma^{\mu}{}_{\nu\rho}$, and hence vanishing Riemann tensor. As $R^{\mu}{}_{\nu\alpha\beta} = 0$ is a tensor, it then must be zero in any coordinate system.

If we know that $R^{\mu}{}_{\nu\alpha\beta} = 0$ everywhere, we say that the space–time is flat. Very often we use the descriptions 'flat' and 'Minkowski' interchangeably.

The Riemann tensor has various symmetries under exchange or permutation of its indices, listed below.

- Antisymmetry on the last two indices (which follows directly from the definition):

$$R^{\mu}{}_{\nu\alpha\beta} = -R^{\mu}{}_{\nu\beta\alpha}. \tag{5.49}$$

- Antisymmetry on the first two indices when they are on the same level:

$$R^{\mu\nu}{}_{\alpha\beta} = -R^{\nu\mu}{}_{\alpha\beta}. \tag{5.50}$$

This is shown in Problem 5.12.

- Cyclic symmetry on the last three indices, sometimes known as the first Bianchi identity:

$$R^{\mu}{}_{\nu\alpha\beta} + R^{\mu}{}_{\alpha\beta\nu} + R^{\mu}{}_{\beta\nu\alpha} = 0. \tag{5.51}$$

This is also shown in Problem 5.12.

- Exchange of pairs of indices (when they are set at the same level):

$$R_{\mu\nu\alpha\beta} = R_{\alpha\beta\mu\nu}.\tag{5.52}$$

A general 4-index tensor in four dimensions has $4^4 = 256$ components. However, because of these symmetry properties, it turns out that the Riemann tensor has only 20 independent components. If space–time has n dimensions, the number of independent components turns out to be $n^2(n^2 - 1)/12$. So in two dimensions there is just one independent component: we will see that this is the intrinsic curvature introduced at the end of Chapter 4.

Finally, the Riemann tensor satisfies a differential identity, known as the second Bianchi identity

$$R^{\mu}{}_{\nu\alpha\beta;\tau} + R^{\mu}{}_{\nu\tau\alpha;\beta} + R^{\mu}{}_{\nu\beta\tau;\alpha} = 0.\tag{5.53}$$

Proof of this identity, and the symmetry on exchange of pairs of indices, will be much easier once we have studied local inertial frames.

5.8 Riemann curvature of the two-sphere

Let's continue with the two-dimensional curved space that we worked out the Christoffel symbols for. Let's start with $R^{\theta}{}_{\theta\theta\theta}$. From its definition,

$$R^{\theta}{}_{\theta\theta\theta} = \Gamma^{\theta}{}_{\theta\theta,\theta} - \Gamma^{\theta}{}_{\theta\theta,\theta} + \Gamma^{\theta}{}_{\theta d}\Gamma^{d}{}_{\theta\theta} - \Gamma^{\theta}{}_{\theta d}\Gamma^{d}{}_{\theta\theta},\tag{5.54}$$

where d takes the values θ, ϕ. We can now see that this equals zero, since every term contains either $\Gamma^{\theta}{}_{\theta\theta}$ or $\Gamma^{\theta}{}_{\theta\phi}$ which are both zero. A more powerful observation though is that the first term cancels the second, and the third one the fourth, without us having to calculate any Christoffel symbols at all. This particular Riemann tensor element vanishes whatever the metric is, as an immediate consequence of the antisymmetry of the Riemann tensor under exchange of the last two indices.

By the same asymmetry argument, Riemann tensor components such as $R^{\phi}{}_{\phi\phi\phi}$ or $R^{\theta}{}_{\theta\phi\phi}$ are also guaranteed to be zero without needing any calculation.

Let us instead consider $R^\theta{}_{\phi\phi\theta}$. From the definition

$$
\begin{aligned}
R^\theta{}_{\phi\phi\theta} &= \partial_\phi \Gamma^\theta{}_{\phi\theta} - \partial_\theta \Gamma^\theta{}_{\phi\phi} + \Gamma^\theta{}_{d\phi} \Gamma^d{}_{\phi\theta} - \Gamma^\theta{}_{d\theta} \Gamma^\rho{}_{\phi\phi} \\
&= \partial_\theta \left(\sin\theta \cos\theta \right) - \cos^2\theta \\
&= \cos^2\theta - \sin^2\theta - \cos^2\theta \\
&= -\sin^2\theta .
\end{aligned}
\tag{5.55}
$$

The non-zero value affirms we are dealing with a genuinely curved space. There is only one independent non-zero value of the Riemann tensor in two dimensions; all others are related to equation (5.56), or zero, by the symmetries of the indices.

If we look at the Riemann tensor with the first two indices raised, and in the same position at both levels, we find

$$
R^{\theta\phi}{}_{\theta\phi} = \frac{1}{a^2}.
\tag{5.56}
$$

This is identical in value to the Gaussian curvature we calculated in equation (4.32). Indeed, provided the metric is diagonal, the Riemann tensor in this rank-$\binom{2}{2}$ form *is* the Gaussian curvature of the sphere.

This also applies in higher-dimensional spaces, in which case we call $R^{\alpha\beta}{}_{\alpha\beta}$ (without a sum on α and β) the **sectional curvature** of the surface defined by varying the coordinates α and β. If the metric is not diagonal, a more complicated formula is needed to obtain the sectional curvature.

5.9   More tensors describing curvature

Having got the Riemann tensor, we can define other curvature tensors from it by summing over the indices. The important ones are the **Ricci tensor**[3]

$$
R_{\mu\nu} \equiv R^\alpha{}_{\mu\alpha\nu} ,
\tag{5.57}
$$

and the **Ricci scalar**,

$$
R \equiv R^\mu{}_\mu \equiv g^{\mu\nu} R_{\mu\nu} ,
\tag{5.58}
$$

which, of course, inherit the proper tensor transformation properties from those of the Riemann tensor.

---

[3] Some authors define the Ricci tensor as the contraction on the first and fourth indices, making it the negative of our version.

Because they are made by summing over a subset of components of the Riemann tensor, they contain less information. The Ricci tensor is symmetric and so has 10 independent components, while the Ricci scalar, being a scalar, has just one component. Incidentally, if you are wondering why it is the first and third indices that are contracted to form the Ricci tensor, Problem 5.13 shows that any other pairing gives either zero or the same result up to a sign, so there really is just one way to form a two-index tensor from the Riemann tensor. As we will see, it is possible for these derived tensors to vanish even when the Riemann tensor does not: the vanishing of the Ricci tensor is no guarantee of a flat space–time.

5.10 Local inertial frames and local flatness

In Chapter 4, we discussed how physical laws would appear in a freely falling laboratory, and concluded that the laws of Special Relativity should apply in sufficiently small regions of space–time. The geometric statement of this conclusion is that there should be a frame, or choice of coordinates, in which the metric approaches the Minkowski metric at a chosen point. Furthermore, one should be able to choose coordinates in which the Christoffel symbols vanish, so that the covariant derivative looks like the ordinary partial derivative of Special Relativity. We call this a **local inertial frame**.

As the Minkowski metric has zero Riemann tensor, it has no curvature. As mentioned previously, we call it and its space–time flat. The property of a general space–time that there is a local inertial frame at every point is sometimes called **local flatness**. This terminology is a little misleading, as space–time cannot be made flat at an arbitrary point, in the sense of its curvature vanishing. Curvature is described by the Riemann tensor, which transforms linearly. If it is non-zero in one set of coordinates, it is non-zero in all.

We will now outline a proof of the existence of local inertial frames. First, we note that $g_{\mu\nu}$ at any point P in space–time can be viewed as a symmetric matrix, with rows and columns labelled by μ and ν. Standard theorems of linear algebra tell us it can be diagonalised with an orthogonal matrix, and further matrix multiplications can be constructed to bring the metric into the Minkowski form $g'_{\mu\nu} = \text{diag}(-1, +1, +1, +1)_{\mu\nu} = \eta_{\mu\nu}$, provided it has one negative and three positive eigenvalues. Let us call the product of these matrices $\Lambda^\alpha{}_\mu$, so that

$$\eta_{\mu\nu} = \Lambda^\alpha{}_\mu \Lambda^\beta{}_\nu g_{\alpha\beta}. \tag{5.59}$$

Hence, coordinates satisfying

$$\frac{\partial x^\alpha}{\partial x'^\mu}\bigg|_P = \Lambda^\alpha{}_\mu \tag{5.60}$$

will bring any metric into Minkowski (sometime called local Lorentz) form at the point P. The sum of the diagonal entries of the metric in this form is called the **signature** of the metric. Only metrics with signature $+2$, referred to as Lorentzian metrics, are relevant for General Relativity.

To show that the Christoffel symbols can also be made to vanish at the same point is a little more involved, and we will just aim to make it plausible rather than give a detailed demonstration. First, we note that if we can make all derivatives of the metric $g'_{\mu\nu,\rho} = 0$ at the point P, we will have accomplished our goal. Then, we write

$$M^{\alpha}{}_{\mu\nu} = \left.\frac{\partial^2 x^{\alpha}}{\partial x'^{\mu}\partial x'^{\nu}}\right|_{P}. \tag{5.61}$$

This object, the second derivative of the inverse coordinate transform $x^{\mu}(x')$ at the point P, has 40 independent components. There are fewer than 4^3 because of the symmetry of the lower two indices.

Differentiating the equation for the transformation of the metric (5.6) and evaluating at point P, we find that in order to make the derivatives of the transformed metric vanish, we require a solution to the equations

$$0 = M^{\alpha}{}_{\mu\rho}\Lambda^{\beta}{}_{\nu}g_{\alpha\beta} + \Lambda^{\alpha}{}_{\mu}M^{\beta}{}_{\nu\rho}g_{\alpha\beta} + \Lambda^{\alpha}{}_{\mu}\Lambda^{\beta}{}_{\nu}g_{\alpha\beta,\rho}. \tag{5.62}$$

If we can solve this equation for $M^{\alpha}{}_{\mu\rho}$, we will have found a coordinate system in which the Christoffel symbols vanish at the point P, and therefore covariant derivatives behave as partial derivatives as desired.

We can plausibly argue that a solution exists on the basis of counting the independent quantities in the equation. Noting that equation (5.62) is symmetric on the indices μ and ν, we see that it represents 40 separate equations. This should be just enough to find the 40 independent quantities $M^{\alpha}{}_{\mu\rho}$. Constructing an explicit expression in terms of the $\Lambda^{\alpha}{}_{\mu}$, $g_{\alpha\beta}$, and $g_{\alpha\beta,\rho}$ is rather time-consuming, and we will not display it.

Finally, we attempt the same argument with the second derivatives of the metric $g_{\mu\nu,\rho,\sigma}$. Given the symmetry in the index pairs $\mu\nu$ and $\rho\sigma$, there are 100 independent quantities to try to remove with a coordinate transformation. However, the next derivative of the coordinate function at the point P,

$$N^{\alpha}{}_{\mu\nu\rho} = \left.\frac{\partial^3 x^{\alpha}}{\partial x'^{\mu}\partial x'^{\nu}\partial x'^{\rho}}\right|_{P} \tag{5.63}$$

has only 80 components. Hence, there is no coordinate transform which can set all the second derivatives of the metric to zero. This implies that the Riemann tensor, which contains second derivatives, cannot be made to vanish by a coordinate transform. This conclusion is consistent with our earlier argument using its tensorial nature.

However, in a local inertial frame, the Riemann tensor does have a simpler form that involves only second derivatives of the metric:

$$R_{\mu\nu\alpha\beta} = \frac{1}{2}\left(g_{\mu\beta,\nu,\alpha} - g_{\nu\beta,\mu,\alpha} - g_{\mu\alpha,\nu,\beta} + g_{\nu\alpha,\mu,\beta}\right). \qquad (5.64)$$

Proof of this equation is part of Problem A3.2. Note that the difference between the number of second derivatives of the metric and the number of available parameters in the coordinate transformation is 20, precisely the number of independent components of the Riemann tensor in 4 space–time dimensions.

The existence of a local inertial frame at every point has another very useful consequence. In this frame, the metric becomes the Minkowski metric, and covariant derivatives become partial derivatives. Conversely, if we can formulate a law of physics in a Special Relativistic form, that is using objects which are tensors under Lorentz transformations, we can immediately promote it to a tensor equation satisfying general coordinate invariance by replacing all occurrences of the Minkowski metric by a general space–time metric $g_{\mu\nu}$ and all the partial derivatives by covariant derivatives. The equation is then compatible with General Relativity.

This might not necessarily lead to a correct law of physics, because one or more of the tensors describing the curvature may need to be introduced, which one would never discover by Special Relativistic arguments alone. Nonetheless, the fact that a tensor equation in Special Relativity can be easily generalised will be very useful to us in Chapter 6.

Problems

5.1 Obtain the three-dimensional metrics corresponding to (a) cylindrical polar coordinates and (b) spherical polar coordinates by transforming from Cartesian coordinates.

5.2* For two-dimensional diagonal metrics, Gauss showed that the curvature K is given by the lengthy formula

$$K = \frac{1}{2g_{11}g_{22}}\left\{-\frac{\partial^2 g_{11}}{\partial(x^2)^2} - \frac{\partial^2 g_{22}}{\partial(x^1)^2} + \frac{1}{2g_{11}}\left[\frac{\partial g_{11}}{\partial x^1}\frac{\partial g_{22}}{\partial x^1} + \left(\frac{\partial g_{11}}{\partial x^2}\right)^2\right] \right.$$
$$\left. + \frac{1}{2g_{22}}\left[\frac{\partial g_{11}}{\partial x^2}\frac{\partial g_{22}}{\partial x^2} + \left(\frac{\partial g_{22}}{\partial x^1}\right)^2\right]\right\}.$$

[Imagine how the four-dimensional curvature formula for non-diagonal metrics might look were it written out like this, which should help you to realise how efficiently our index notation deals with such expressions.] Compute the curvature for the metric

$$ds^2 = f(\theta)\,dr^2 + d\theta^2,$$

where $f(\theta)$ is an arbitrary function. Find a non-constant function $f(\theta)$ which gives zero curvature everywhere.

5.3   Suppose that in Problem 5.2, the metric had instead been

$$ds^2 = f(r)\,dr^2 + d\theta^2.$$

What would the curvature be in that case? Explain why it is not necessary to evaluate K in this case.

5.4   From equation (5.35), show that the equation for parallel transport of a vector **v** along a line of latitude (constant θ) is

$$\frac{\partial v^\theta}{\partial \phi} - \sin\theta\,\cos\theta\,v^\phi \;=\; 0;$$

$$\frac{\partial v^\phi}{\partial \phi} + \frac{\cos\theta}{\sin\theta}v^\theta \;=\; 0,$$

Solve these equations to show that a vector with components (v_0^θ, v_0^ϕ) at $\theta = \theta_0$, $\phi = 0$, becomes

$$v^\theta(\phi) \;=\; v_0^\theta \cos[(\cos\theta_0)\phi] + v_0^\phi \sin\theta_0 \sin[(\cos\theta_0)\phi];$$

$$v^\phi(\phi) \;=\; v_0^\phi \cos[(\cos\theta_0)\phi] - \frac{v_0^\theta}{\sin\theta_0}\sin[(\cos\theta_0)\phi].$$

as it is parallel transported around the line of latitude. Check that **v** has constant magnitude.

5.5*   If a^μ and b^ν are contravariant 4-vectors, show that covariant differentiation obeys the product rule

$$(a^\mu b^\nu)_{;\rho} = a^\mu{}_{;\rho}\, b^\nu + a^\mu\, b^\nu{}_{;\rho}.$$

5.6   Show that for a scalar quantity ϕ, the order in which covariant derivatives are taken doesn't matter, that is,

$$\phi_{,\alpha;\beta} = \phi_{,\beta;\alpha} \, .$$

5.7*   Use the assumed symmetry of the Christoffel symbols $\Gamma^\tau_{\mu\nu} = \Gamma^\tau_{\nu\mu}$, and the metric-compatibility condition $g_{\mu\nu;\rho} = 0$, to show that the Christoffel symbols are

$$\Gamma^\tau_{\mu\nu} = \frac{1}{2} g^{\tau\sigma} \left(g_{\sigma\mu,\nu} + g_{\sigma\nu,\mu} - g_{\mu\nu,\sigma} \right) \, .$$

[Hint: write out the metric-compatibility condition three times, with indices permuted, and look for a useful linear combination.]

5.8   [Hard!] Show that the transformation law of the Christoffel symbols must be

$$\Gamma'^\mu_{\nu\rho} = \left[\left(\frac{\partial x'^\mu}{\partial x^\sigma} \right) \Gamma^\sigma_{\kappa\lambda} - \left(\frac{\partial^2 x'^\mu}{\partial x^\kappa \partial x^\lambda} \right) \right] \left(\frac{\partial x^\kappa}{\partial x'^\nu} \right) \left(\frac{\partial x^\lambda}{\partial x'^\rho} \right) ,$$

in order that the covariant differential of a 4-vector transforms as a 4-vector. Check that the expression (5.12) in terms of derivatives of the metric transforms in this way.

5.9*   Consider the expression $(a^\mu b_\mu)_{;\rho}$, where a^μ is an arbitrary vector. Show that the formula for covariant differentiation of a covariant 4-vector must be

$$b_{\mu;\rho} = b_{\mu,\rho} - \Gamma^\nu_{\mu\rho} b_\nu \, .$$

5.10*   Recall that the covariant derivatives of 4-vectors are defined by

$$a_{\mu;\sigma} = a_{\mu,\sigma} - \Gamma^\nu_{\mu\sigma} a_\nu ; \quad a^\mu_{;\sigma} = a^\mu_{,\sigma} + \Gamma^\mu_{\sigma\nu} a^\nu \, ,$$

and the Riemann tensor by

$$a^\mu_{;\alpha;\beta} - a^\mu_{;\beta;\alpha} = -R^\mu_{\nu\alpha\beta} a^\nu \, .$$

Use these to prove that

$$R^\mu_{\ \nu\alpha\beta} = \Gamma^\mu_{\ \beta\nu,\alpha} - \Gamma^\mu_{\ \alpha\nu,\beta} + \Gamma^\mu_{\ \alpha\rho}\Gamma^\rho_{\ \beta\nu} - \Gamma^\mu_{\ \beta\rho}\Gamma^\rho_{\ \alpha\nu}.$$

5.11

a) Consider a general two-dimensional space with coordinates θ and ϕ. If you were asked to compute a particular component of the Ricci tensor, namely $R_{\theta\theta}$, what would be the minimum number of components of the Riemann tensor required to do that, and which would they be?

b) Now specialise to a two-dimensional sphere, which has metric $ds^2 = r^2 d\theta^2 + r^2 \sin^2\theta\, d\phi^2$, where r is a constant. Use the Christoffel symbols for this metric given within this chapter to compute $R_{\theta\theta}$ for the two-sphere.

c) Given that $R_{\phi\phi} = \sin^2\theta$, compute the Ricci scalar $R \equiv R^\mu_{\ \mu}$.

5.12

a) Consider the scalar quantity $\phi = a^\rho b_\rho$. Use the property that the order of covariant derivatives of a scalar is irrelevant (Problem 5.6) to verify equation (5.48).

b) By raising the vector index on the covariant vector, demonstrate the anti-symmetry of the Riemann tensor on the first two indices, equation (5.50).

c) Apply definition (5.19) of covariant differentiation to the vector $a_\mu = \psi_{;\mu}$, where ψ is a scalar, and consider cyclic permutations on the indices μ, α, β. Hence, prove the identity (5.51).

5.13* The Ricci tensor is defined by contracting the first and third indices of the Riemann tensor; $R_{\mu\nu} \equiv R^\alpha_{\ \mu\alpha\nu}$, rather than any other pair. Demonstrate from the symmetry properties of the Riemann tensor that all of the possible definitions give a tensor which is either identically zero or equal to $\pm R_{\mu\nu}$.

5.14* Prove the second Bianchi identity, given in equation (5.53):

$$R^\mu_{\ \nu\alpha\beta;\tau} + R^\mu_{\ \nu\tau\alpha;\beta} + R^\mu_{\ \nu\beta\tau;\alpha} = 0.$$

[Hint: specialise to a local inertial frame.]

Chapter 6

Describing Matter

How does matter make space–time curve? · the energy–momentum tensor is the key

Having decided that gravitational effects can be attributed to curvature of space–time, we have set ourselves two tasks. The first is to understand how physical laws will operate in curved space–time, while ensuring no violations of well-established laboratory results across the whole spectrum of physical phenomena. Here we will be guided by the Correspondence Principle. The second is to develop an equation which governs how matter, the source of gravity, gives rise to the space–time curvature. In each case, the appropriate framework is tensor calculus, which ensures outcomes that are independent of the choice of coordinates.

6.1   The Correspondence Principle

Often, physical laws take the form of partial differential equations, as normally they express how physical quantities vary with respect to space and time. An archetype is Maxwell's equations which govern the space and time dependence of electric and magnetic fields in the presence of sources, and which are readily expressible in a tensor notation.

The Correspondence Principle, introduced in Section 4.5, advises us that we should seek to recover the conventional physical laws in the limit where gravity vanishes, i.e. where the space–time metric tends to the flat Minkowski form. We can enforce this by converting flat space–time laws to curved space–time laws as follows:

- The Minkowski metric $\eta_{\mu\nu}$ is replaced with the space–time metric $g_{\mu\nu}$.

Introducing General Relativity, First Edition. Mark Hindmarsh and Andrew Liddle.
© 2022 John Wiley & Sons Ltd. Published 2022 by John Wiley & Sons Ltd.
Companion website: www.wiley.com/go/hindmarsh/introducingGR

- Partial derivatives are replaced by covariant derivatives (colloquially, commas are replaced by semi-colons).

This simple prescription ensures that our equations remain properly formulated tensor equations, which in curved space–time requires the use of the covariant derivative. Since in the flat space–time limit partial and covariant derivatives become identical, due to vanishing of the Christoffel symbols, we are guaranteed to recover the original physical laws. Hence, we can be sure that we are not harming the validity of any physical laws derived in circumstances where gravity is negligible.

This procedure implies that physics in a freely falling frame always behaves as if gravity were absent. This leads to the Strong Equivalence Principle. However, as we noted at the end of Chapter 5, the procedure is not always unique.

One further expectation from the Correspondence Principle is that we should recover Newtonian gravitational theory in situations where gravity has no relativistic sources. This is very different from the simple prescription above for non-gravitational laws, and indeed, it is far from trivial that Einstein's General Relativity does approximate Newtonian gravity in this limit, as we will later see.

6.2 The energy–momentum tensor

6.2.1 General properties

What law governs the space–time curvature itself? We know that matter is to be responsible for gravitational forces, and having decided that gravity is due to curvature, we need a formula which takes the schematic form

$$\text{curvature} \quad \longleftrightarrow \quad \text{matter}. \tag{6.1}$$

As in the discussion above, only tensors are allowed if the theory is to be covariant, and hence any differentiations have to be covariant derivatives. Additionally, the Correspondence Principle demands that we must recover Poisson's equation for the Newtonian gravitational potential Φ in the weak field slow-motion limit

$$\nabla^2\Phi = 4\pi G\rho\,, \tag{6.2}$$

where ρ is the density of matter.

Hence, we need a tensor which describes the properties of matter, and in particular contains the density. This tensor must also incorporate conservation laws obeyed by matter, such as conservation of mass–energy and momentum. In keeping with philosophy of Section 6.1, we will initially place our discussion in the

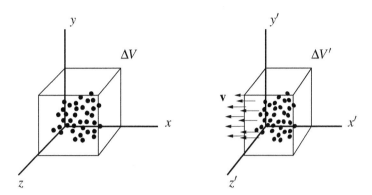

Figure 6.1 Left: ΔN stationary particles in a volume ΔV, with average density $n = \Delta N/\Delta V$. Right: The same particles as viewed from a frame moving in the $+x$ direction with speed v. In the second frame, the volume $\Delta V'$ is reduced, as the dimension parallel to the velocity is Lorentz contracted. The average density of particles is therefore greater in the second frame, and there is a flux of particles in the $-x$ direction.

context of Special Relativity, ignoring gravity, and then use the Correspondence Principle to extend it to curved space–time.

6.2.2 Conservation laws and 4-vector flux

To start the consideration of what kind of tensor might contain the density, let us suppose that the matter is comprised of a large number of very slowly-moving particles of mass m distributed over some volume Ω, and that we can neglect any forces between them. Let us also suppose that the particles can be neither created nor destroyed, so that the total particle number N is constant. We will construct a tensor under these assumptions, which we can then generalise.

We can define an average particle number density n over a small region $\Delta V(\mathbf{x})$ containing ΔN of the particles through $n \equiv \Delta N/\Delta V$, which will be a smooth function of the position of the centre of the region $\mathbf{x}$ if the number of particles is large enough.

What is the particle density seen by an observer moving with speed v in, say, the $+x$-direction (see Figure 6.1)? The observers must agree on the number of particles in the volume ΔN, but the Lorentz–Fitzgerald contraction in the x direction will make the volume according to the second observer $\Delta V'$ appear smaller, according to $\Delta V' = \Delta V/\gamma$. Hence, the density for the second observer is

$$n' = \gamma n. \tag{6.3}$$

This looks like part of the transformation law for the zeroth component of a vector.

At the same time, the second observer now sees a collective motion of the particles in the $-x$ direction, with flux (number of particles crossing unit area per second)

$$j'^i = -vn'\hat{x}^i, \tag{6.4}$$

where $\hat{x}^i$ are the components of a unit vector in the x direction. This looks like the spatial part of a Lorentz transformation of a 4-vector. A further transformation to another moving frame can be performed to convince oneself that the particle density and flux can indeed be assembled into a flux 4-vector

$$j^\mu = (cn, j^i). \tag{6.5}$$

The conservation of particle number is encoded in the equation, normally called the **continuity equation**

$$\partial_\mu j^\mu = \dot{n} + \frac{\partial j^i}{\partial x^i} = 0. \tag{6.6}$$

If we integrate the continuity equation over a volume V, we see that the rate of change of the number of particles in that volume N_V is

$$\dot{N}_V = \int_V d^3x \dot{n} = -\int_V d^3x \partial_i j^i. \tag{6.7}$$

Using the divergence theorem,

$$\dot{N}_V = -\int_{\partial V} d^2 S_i j^i, \tag{6.8}$$

where ∂V is the boundary of the volume V, with outward-directed surface element $d^2 S_i$. Hence, the particle number in a volume changes according to the total flux through its boundary. If the volume V is taken to include the whole volume Ω, so that the flux at the boundary drops to zero, we recover the expression of the conservation of particle number

$$\dot{N} = 0. \tag{6.9}$$

We can conclude that for every conserved quantity like N, we expect to be able to find a 4-vector, with the density n as a zeroth component, which obeys a continuity equation like equation (6.6).

6.2.3 Energy and momentum belong in a rank-2 tensor

Now let us consider the mass density of the particles in the original frame, $\rho = mn$, and ask how it looks in the moving frame. Each of the particles now has mass $m' = \gamma m$, as described in Section 3.3, and hence the mass density in the moving frame is

$$\rho' = \gamma^2 \rho. \tag{6.10}$$

The two powers of γ, one from the increased mass and one from the decreased volume, are an immediate sign that the energy density does not belong in a 4-vector: instead, they point to its belonging in a rank-2 tensor. We also note that the particles in the moving frame now have momentum per unit volume

$$\pi'^i = -m'vn'\hat{x}^i = -\gamma^2 mvn\hat{x}^i. \tag{6.11}$$

Hence, we expect this rank-2 tensor to also contain the momentum density.

Indeed, our argument that every conserved quantity should be associated with a 4-vector current leads us to a similar conclusion. The total mass–energy and momentum

$$E = \int \rho c^2 d^3x; \quad p^i = \int \pi^i d^3x, \tag{6.12}$$

are both conserved, and they are themselves the components of a momentum 4-vector $p^\mu = (E/c, p^i)$. Hence, there should be four sets of 4-vector fluxes of the form

$$j_E^\mu; \quad j_{p^i}^\mu, \tag{6.13}$$

satisfying continuity equations, which transform into each other as 4-vectors. These two fluxes can be assembled into an object with two indices, $T^{\mu\nu}$, with

$$T^{\mu 0} = j_E^\mu; \quad T^{\mu i} = j_{p^i}^\mu. \tag{6.14}$$

The total energy and momentum in a particular frame are recovered by integration over the spatial coordinates in that frame,

$$E = \int T^{00} d^3x; \quad p^i = \frac{1}{c} \int T^{i0} d^3x, \tag{6.15}$$

where the factor of $1/c$ keeps the dimensions of the components consistent. The rank-2 tensor $T^{\mu\nu}$ is the **energy–momentum tensor** or **stress–energy tensor**. In the Special Relativistic argument presented above, energy–momentum conservation has a particularly elegant expression in terms of a set of four continuity equations,

$$T^{\mu\nu}{}_{,\nu} = 0. \qquad (6.16)$$

When $\mu = 0$, we have the continuity equation for energy conservation, while the remaining three values of the μ index give momentum conservation.

6.2.4 Symmetry of the energy–momentum tensor

Not all the components of $T^{\mu\nu}$ are independent. We have already seen that T^{i0} is the momentum density. By construction, T^{0j} is the flux of mass–energy in the j-th direction, meaning the rate per unit volume at which mass–energy crosses a surface oriented in the direction $\hat{x}^j$. But this is just the same thing as momentum density. Hence, we conclude that $T^{0i} = T^{i0}$.

The spatial components T^{ij} contain the flux of momentum π^i across a surface oriented in the direction $\hat{x}^j$. In particular, consider two small adjacent cubes with volume ΔV at $\mathbf{x}_0$ and $\mathbf{x}_1$ separated by a surface ΔS^{01}_j. A flux of momentum across an area results in a force, by Newton's second law. Hence, the first volume element exerts a force on the second of

$$\Delta F^i_{01} = T^{ij}(\mathbf{x}_0)\Delta S^{01}_j , \qquad (6.17)$$

with the second returning a force of

$$\Delta F^i_{10} = -T^{ij}(\mathbf{x}_1)\Delta S^{01}_j . \qquad (6.18)$$

Figure 6.2 shows a force diagram. By considering the net torque on the first element from all six of the surrounding elements, it can be shown that the T^{ij} is symmetric: otherwise, the net torque on an element would diverge as the size is taken to zero. Hence, the energy–momentum tensor as a whole is symmetric:

$$T^{\mu\nu} = T^{\nu\mu}. \qquad (6.19)$$

There are therefore only 10 independent components, which are further related by the equations of continuity.

6.2.5 Energy–momentum of perfect fluids

We now introduce the important concept of a perfect fluid, also called an ideal fluid. We suppose it to be made of a huge number of particles, such that all the

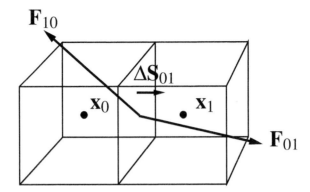

Figure 6.2 Forces exerted by adjacent volume elements on each other, when non-zero energy–momentum is present. For example, $\mathbf{F}_{01}$ is the force exerted by the element at $\mathbf{x}_0$ on the element at $\mathbf{x}_1$. The forces depend on the components of energy–momentum tensor T^{ij} at the centres of the volume elements according to equations (6.17) and (6.18). They can be taken to act through the centre of the surface element separating them, whose normal is denoted $\Delta \mathbf{S}^{01}$.

components of the energy–momentum tensor can be treated as smooth functions of time and space coordinates (t, x^i). The particles may have some internal motion, but collide frequently, interchanging energy and momentum, and the mean free path is much shorter than any length scale we will need to consider. We can locally find a frame in which the momentum density is zero. In this frame, the local fluid 3-velocity, defined as

$$v^i = \frac{T^{i0}}{T^{00}}, \tag{6.20}$$

vanishes. This is the rest frame of the fluid.

In a perfect fluid at rest, the only force that can be exerted is the pressure, which is always perpendicular to a surface. Hence, for the adjacent volume elements considered above, ΔF_{01}^i must be parallel to ΔS_j^{01} and equal to the pressure multiplied by the area. Hence, the energy–momentum tensor of a perfect fluid at rest is diagonal, with components equal to the pressure,

$$T^{ij} = \delta^{ij} p. \tag{6.21}$$

This supplements our original understanding that the 00 component is the energy density,

$$T^{00} = \rho c^2. \tag{6.22}$$

By performing a Lorentz transformation to a frame moving with 4-velocity u^μ, one finds that the energy–momentum tensor for a moving perfect fluid is (see Problem 6.1)

$$T^{\mu\nu} = \left(\rho + \frac{p}{c^2}\right) u^\mu u^\nu + \eta^{\mu\nu} p. \tag{6.23}$$

A fluid is called **barotropic** if the pressure is a function of the energy density only, $p = p(\rho)$. The functional relationship between the pressure and other thermodynamic variables of the system is known as the **equation of state**.

There are some important special cases for perfect barotropic fluids.

- **Dust or matter:** This refers to our original model of slowly moving particles with no forces between them. The only non-zero component of the energy–momentum tensor in the local rest frame is T^{00}, so the equation of state is the trivial relation $p = 0$.

- **Radiation:** This is the special case of a fluid where the equation of state is $p = \rho c^2/3$, which represents a gas of ultra-relativistic particles. In particular, it describes a gas consisting of massless particles, photons being a key example. If the photons occupy a thermal distribution at temperature T, the rest-frame density ρ is proportional to T^4

- **Vacuum energy:** It is possible for space which is empty of particles to have a non-zero energy–momentum tensor

$$T^{\mu\nu} = -V_0 \eta^{\mu\nu}, \tag{6.24}$$

where V_0 is a constant. In classical physics, this sounds rather nonsensical, but quantum mechanics has taught us of the unavoidable zero-point energy of quantum fields even when no particles are present, and if such an energy exists, we should expect it to contribute to gravitational effects. This term is equivalent to a perfect fluid with equation of state $p = -\rho c^2$, i.e. a fluid with negative pressure $p = -V_0$, and is closely related to the cosmological constant Λ introduced in Chapter 7. Note that because by definition $\eta^{\mu\nu}$ is unchanged by Lorentz transformation, all observers agree on this energy–momentum tensor, as is necessary if it is to be interpreted as being empty space. The vacuum is said to be Lorentz invariant.

Real fluids show departures from the perfect fluid properties. For example, they can support forces which are not perpendicular to any given surface, known as shear stresses, which cause planes of the fluid to resist slipping past one another. This is a sign of viscosity in the fluid, which shows up as extra off-diagonal components in the energy–momentum tensor, not accounted for in equation (6.23).

Real fluids also conduct heat, meaning that energy can flow from place to place even without the fluid moving. In a perfect fluid, energy is transported only by fluid motion.

6.2.6 The energy–momentum tensor in curved space–time

The final step is to use our understanding of local inertial frames to show how the Special Relativistic form of the energy–momentum tensor can be adapted for General Relativity. As discussed in Section 5.10, the way to enhance a tensor expression so that it is generally covariant (and so satisfies the principle of General Relativity) is to replace partial derivatives by covariant derivatives and the Minkowski metric $\eta_{\mu\nu}$ by the metric tensor of the full space–time $g_{\mu\nu}(x)$. Objects which are tensors under Lorentz transformations, such as the 4-velocity, are guaranteed to become tensors under general coordinate transformations.

The energy–momentum conservation law (6.16) becomes

$$T^{\mu\nu}{}_{;\nu} = 0. \tag{6.25}$$

We call this general-relativistic equation the covariant energy–momentum conservation law, even though energy and momentum are no longer necessarily conserved in the same way as in Special Relativity. The understanding of the sense in which energy–momentum is conserved in General Relativity is the achievement of the mathematician Emmy Noether. Her work uncovered a profound connection between conservation laws and invariances, in this case, the invariance of physical quantities under coordinate changes. This connection still underpins much of modern theoretical physics.

As an example of the procedure of moving from Special to General Relativity, we can immediately write down the energy–momentum tensor for a perfect fluid in a general space–time,

$$T^{\mu\nu} = \left(\rho + \frac{p}{c^2}\right) u^{\mu} u^{\nu} + g^{\mu\nu} p. \tag{6.26}$$

Now the metric components are no longer just ± 1 or 0, which means that the physical quantities like energy density and pressure are most directly related to the mixed tensor, $T^{\mu}{}_{\nu}$. For a stationary perfect fluid, we have

$$T^0{}_0 = -\rho c^2, \quad T^i{}_j = p\delta^i{}_j. \tag{6.27}$$

These relationships hold for a perfect fluid in any space–time, whereas equations (6.21) and (6.22) apply only in the space–time of the Special Relativity. Problem 6.6 explores the consequences of the covariant conservation of a perfect fluid.

Problems

6.1* The aim of this problem is to verify that the energy–momentum tensor for a perfect fluid is indeed equation (6.23).

a) Consider the matrix

$$\Lambda^{\mu}{}_{\nu} = \begin{pmatrix} \gamma & \gamma v_j/c \\ \gamma v^i/c & \delta^i_j + (\gamma-1)\hat{v}^i\hat{v}_j \end{pmatrix},$$

where μ takes values $(0,i)$, ν takes values $(0,j)$, v^i are the components of a vector $\mathbf{v}$, $\gamma = 1/\sqrt{1-\mathbf{v}^2/c^2}$, and $\hat{v}^i$ are the components of the unit vector in the direction of $\mathbf{v}$. Apply this matrix to the coordinate 4-vector x^{ν}, and show that the result can be interpreted as a Lorentz boost to a frame moving with speed v in the direction $-\hat{\mathbf{v}}$.

b) Apply this Lorentz transformation to the energy–momentum tensor of the stationary perfect fluid, equations (6.21) and (6.22), and verify that the result can be re-organised into the given form (6.23), consistent with the interpretation of a fluid moving with velocity $\mathbf{v}$ in the new frame. [You may need to remind yourself of the subtleties of treating Lorentz transformations as matrix multiplications, discussed in Chapter 3.]

6.2 Show that the trace of the energy–momentum tensor $T^{\mu}{}_{\mu}$ vanishes for a perfect radiation fluid.

6.3 Show that u^{μ} is an eigenvector of the perfect fluid energy–momentum tensor $T^{\mu}{}_{\nu}$ viewed as a matrix, and find the corresponding eigenvalue. What is the eigenvalue of any 4-vector w^{μ} orthogonal to u^{μ} (i.e. satisfying $w \cdot u = 0$)?

6.4 Explain why $-T^0{}_0 - \rho c^2$ is the kinetic energy density of a perfect fluid in Minkowski space–time. Find the expression in terms of ρ, p, and the speed v. Show that $T^i{}_i - 3p$ is also equal to the kinetic energy density.

6.5 [Hard!] Suppose a perfect fluid with energy density ρ and pressure p is moving non-relativistically with 3-velocity v^i.

a) Write out the components of $T^{\mu\nu}$ in an expansion in v/c, keeping only the first term in each component. Show that the energy–momentum conservation equation $T^{\mu\nu}{}_{,\nu} = 0$ implies

$$\frac{\partial\rho}{\partial t} + \partial_j(wv^j) = 0;$$

$$\frac{\partial}{\partial t}(wv^i) + \partial_j(wu^iu^j) + \partial^i p = 0,$$

where $w = \rho c^2 + p$. [This is sometimes known as the enthalpy density.]

b) Writing the density as $\rho = \bar{\rho}(1+\delta)$, with $\bar{\rho}$ a constant mean density, show that a small density perturbation δ satisfies

$$\left(\frac{\partial^2}{\partial t^2} - c_s^2 \nabla^2\right)\delta = 0,$$

and find c_s^2.

c) Verify that the resulting wave solutions have longitudinal velocity perturbations (i.e. that the velocity is in the same direction as the wave vector). These are therefore sound waves. What is the speed of sound in a perfect radiation fluid?

6.6* [Hard!] In this problem, you show how to derive the Euler equation for a fluid from the energy–momentum conservation equation of General Relativity.

a) Write out the covariant conservation equation in the form $T^{\mu\nu}{}_{;\nu} = 0$ using partial derivatives and Christoffel symbols.

b) In the Newtonian limit (a weak slowly-changing gravitational field induced by non-relativistic sources), we will see in Chapter 7 that the space–time interval may be written as

$$ds^2 = -\left(1 + \frac{2\Phi}{c^2}\right)dt^2 + \left(1 - \frac{2}{\Phi/c^2}\right)(dx^2 + dy^2 + dz^2),$$

where Φ is the Newtonian potential. Show that the non-vanishing Christoffel symbols are, to first order in Φ/c^2,

$$\Gamma^m{}_{00} = \frac{1}{c^2}\Phi^{,m}, \quad \Gamma^0{}_{r0} = \Gamma^0{}_{0r} = \frac{1}{c^2}\Phi_{,r},$$

$$\Gamma^m{}_{nr} = -\frac{1}{c^2}\left(\delta^m_n\Phi_{,r} + \delta^m_r\Phi_{,n} - \delta_{nr}\Phi^{,m}\right).$$

Hence, show that for a non-relativistic fluid the equations in found in part

a) above become, to the same order in v/c and Φ/c^2,

$$\frac{\partial}{\partial t}v^i + (v^j\partial_j)v^i + \frac{\partial^i p}{\rho} + \partial^i\Phi = 0.$$

This is the Euler equation, describing non-relativistic fluid flow in a gravitational field.

Chapter 7

The Einstein Equation

Put it all together and we have the Einstein equation · can we get back to Newton's gravity? · introducing the cosmological constant

The picture of space–time we have developed in Chapters 4 and 5 has led us to a particular set of mathematical ideas which belong to a structure known as Riemannian geometry. In this structure, space–time is an object called a manifold, which is a collection of points that can be assigned real numbers as coordinates, and which is additionally equipped with a metric which gives the distance between points. Riemannian geometry then lets us define important objects such as the Levi-Civita connection and the curvature. The physical content of General Relativity is a relationship between the objects describing the space–time manifold and the matter in the space–time.

By now, we know the general form of the equation that we are after. It should relate the curvature of space–time to the properties of matter, so that gravitational effects can be attributed to a curvature of space–time that is induced by the presence of matter. It must only feature tensor quantities in order to implement general covariance. Ultimately, it should approximate Newton's theory of gravity in the appropriate limits. Following Einstein, we will arrive at a suitable equation by guesswork and then test its viability.

7.1 The form of the Einstein equation

First, we summarise the tensors describing the space–time and the properties of the matter that lives within the space–time. They are

Space–time: $g_{\mu\nu}, R^{\mu}{}_{\nu\alpha\beta}, R_{\mu\nu}, R$

Matter: $T^{\mu\nu}$, which should be covariantly conserved, $T^{\mu\nu}{}_{;\nu} = 0$

Introducing General Relativity, First Edition. Mark Hindmarsh and Andrew Liddle.
© 2022 John Wiley & Sons Ltd. Published 2022 by John Wiley & Sons Ltd.
Companion website: www.wiley.com/go/hindmarsh/introducingGR

How can we put them together?

The first conclusion we can reach is that we cannot use the Riemann tensor alone. There's a simple mathematical reason for this, which is that the Riemann tensor is a four-index tensor, while the energy–momentum tensor is rank two. As those objects transform differently, they cannot be equated to each other. We could try to salvage this by squaring the energy–momentum tensor, but the success of Newton's theory tells us it is mass, not mass-squared, that sources gravity, so this won't work either.

However, it is also instructive to notice a physical reason to avoid the Riemann tensor. In empty space, we have $T^{\mu\nu} = 0$. If the Riemann tensor were in an equation relating it to $T^{\mu\nu}$ it would equal zero wherever there was no matter, implying flat space and hence no gravity. But we know gravity from nearby objects does exist in empty space, for instance the effect of the Sun on the Earth. Therefore, whatever equation we derive must allow some Riemann tensor components to be non-zero even at locations where $T^{\mu\nu}$ is zero.

This encourages us to instead focus on the Ricci tensor, which has the same number of indices as the energy–momentum tensor. Indeed, Einstein's first attempts focussed on the very tempting form $R^{\mu\nu} = \text{constant} \times T^{\mu\nu}$. However, the covariant derivative of the Ricci tensor, $R^{\mu\nu}_{;\nu}$, does not vanish in general, while that of $T^{\mu\nu}$ does. Hence, they cannot be proportional and so this simple form fails.

We need to find a rank-2 tensor which describes the curvature *and* is covariantly conserved. The simplest candidate that we can assemble from the ingredients we have is $R^{\mu\nu} + Cg^{\mu\nu}R$, where C is a constant. It turns out that this is covariantly conserved if and only if $C = -1/2$ (Problem 7.2). We conclude that an appropriate covariantly conserved tensor is

$$G^{\mu\nu} \equiv R^{\mu\nu} - \frac{1}{2}g^{\mu\nu}R. \tag{7.1}$$

This is known as the **Einstein tensor**.

Hence, the simplest equation which relates the curvature of space–time to the properties of matter in a consistent way is

$$G^{\mu\nu} = \kappa T^{\mu\nu}, \tag{7.2}$$

where κ is a constant to be fixed by comparison with observation. This is the **Einstein equation**. It has all the properties we seek, and defines the General Theory of Relativity.

7.2 Properties of the Einstein equation

Here we summarise the main properties of the Einstein equation:

1. Both sides of the Einstein equation feature a symmetric rank-2 tensor. It therefore corresponds to ten equations in all.

2. The aim of the Einstein equation is to determine the ten components of the metric tensor, so in general ten equations are needed.

3. If we are in a physical situation with symmetries, there may be redundancies amongst the equations. For example, if $g_{\mu\nu}$ is diagonal, only four components need to be determined so there will be at most four independent Einstein equations. Properties such as spherical symmetry can reduce this further.

4. Energy–momentum conservation is automatically encoded in the Einstein equations and so we do not need any additional equations to enforce it. However, as in elementary kinematics, it is often useful to use some of the four energy–momentum conservation equations in place of some of the Einstein equations. They are often simpler, for instance being first-order rather than second-order differential equations hence corresponding to first integrals of the Einstein equations.

5. They are coupled nonlinear differential equations. They are extraordinarily hard to solve except under the simplest of physical conditions.

6. A subtle point is that observable consequences of the equations are supposed to be independent of coordinates. We can transform any one of the four coordinates. In practice, this means that four Einstein equations are equivalent to coordinate transformations, leaving at most six independent ones even in the most general circumstances.

7. The Correspondence Principle demands that Newton's theory of gravity be obtained for weak gravitational fields. This fixes the constant, giving $\kappa = 8\pi G/c^4$, as we will see in Section 7.3.

7.3 The Newtonian limit

The Newtonian limit applies in the case where gravitational fields are weak and speeds of sources are much less than the speed of light, so that the field changes only slowly. This means that the $T^{00} = \rho c^2$ component of the energy–momentum tensor is the largest, so Poisson's equation

$$\nabla^2 \Phi = 4\pi G\rho \,, \tag{7.3}$$

is expected to emerge in this limit, from the 00 Einstein equation.

A weak gravitational field means that the metric tensor is close to that of flat space–time, and so can be written as

$$g_{\mu\nu} = \eta_{\mu\nu} + h_{\mu\nu} \,, \tag{7.4}$$

with $|h_{\mu\nu}| \ll 1$ for all components. In Advanced Topic A3, it is shown that substituting this ansatz into the Einstein equation implies that the metric perturbation satisfies (in a suitably chosen set of coordinates)

$$\Box h^{\mu\nu} = -2\kappa \left(T^{\mu\nu} - \frac{1}{2}\eta^{\mu\nu}T \right) \,, \tag{7.5}$$

where $T \equiv T^{\mu}{}_{\mu}$ is the trace of the energy–momentum tensor.

For a stationary perfect fluid, we therefore find that

$$\Box h^{00} = -\kappa(c^2\rho + 3p) \,, \tag{7.6}$$

The assumption of small speeds means that $\partial/\partial t$ is of the order of $v\partial/\partial x$ and hence the time derivatives in $\Box$ can be ignored. For a fluid made of non-relativistic particles, the pressure can be ignored in comparison to the energy density, yielding

$$\nabla^2 h^{00} \simeq -\kappa c^2\rho \,. \tag{7.7}$$

We argued in Section 4.6 that the slow running of clocks in a gravitational field means that $g_{00} \simeq -(1 + 2\Phi/c^2)$; hence, h_{00} can be identified with $-2\Phi/c^2$. In the weak-field limit, $h^{00} = h_{00}$. Therefore, comparing to the Poisson equation, we see that the constant κ must be

$$\kappa = \frac{8\pi G}{c^4} \,. \tag{7.8}$$

One can also derive the weak-field equation for the spatial parts of the metric, which for non-relativistic source is

$$\nabla^2 h^{ij} \simeq -\kappa \delta^{ij} c^2\rho \,. \tag{7.9}$$

Hence, the space–time interval in a weak gravitational field produced by non-relativistic sources can be written

$$ds^2 = -\left(1 + \frac{2\Phi}{c^2}\right) dt^2 + \left(1 - \frac{2\Phi}{c^2}\right) \left(dx^2 + dy^2 + dz^2\right), \qquad (7.10)$$

with Φ obeying the Poisson equation.

In Problem 7.4, you can verify that the Einstein equation with this metric does indeed reproduce the Poisson equation from its 00 component.

7.4 The cosmological constant

In fact, the Einstein tensor is not the most general tensor which has vanishing co-variant derivative. Since the covariant derivative of the metric tensor vanishes, we can add it in with an arbitrary coefficient, giving a slightly more general equation with one extra parameter

$$G^{\mu\nu} + \Lambda g^{\mu\nu} = \frac{8\pi G}{c^4} T^{\mu\nu}. \qquad (7.11)$$

The constant Λ is known as the cosmological constant, whose value is undetermined by theoretical considerations. Einstein introduced it to allow static cosmological solutions, later retracting this once it was established that the Universe was expanding.[1] However, the cosmological constant has resurfaced at intervals ever since, and indeed, at present is being invoked by cosmologists to explain observations indicating that the expansion of the Universe is accelerating.

The cosmological constant is formally equivalent to the vacuum energy introduced in Section 6.2.5; comparing equations (6.24) (with $\eta^{\mu\nu}$ replaced by $g^{\mu\nu}$ to include space–time curvature) and (7.11), we see that the Λ term has the same form as a vacuum energy provided $\Lambda = 8\pi G V_0/c^4$. This gives it the interpretation of being the energy–momentum of empty space. It is a matter of choice whether to include the effects of the vacuum as a separate Λ term or as part of the energy–momentum tensor; in the latter case, it acts as a perfect fluid with energy density and pressure given by

$$\rho c^2 = \frac{\Lambda c^4}{8\pi G}, \qquad p = -\frac{\Lambda c^4}{8\pi G}. \qquad (7.12)$$

For the majority of this book, we will ignore the cosmological constant, looking at its effects only occasionally and otherwise setting it to zero. We will return

---

[1] It is often reported that Einstein said this was his 'biggest blunder'. In fact, this phrase comes from cosmologist George Gamow's autobiography, and it is unclear that Einstein himself ever said it.

to it in our final discussions on cosmology, where it is a key part of the standard cosmological model.

7.5   The vacuum Einstein equation

One of the most important applications of Einstein's equations is to empty space. This will tell us the effects of gravity outside gravitating bodies. Empty space has $T^{\mu\nu} = 0$, giving an Einstein equation

$$R^{\mu\nu} - \frac{1}{2} g^{\mu\nu} R = 0, \tag{7.13}$$

where we neglect any cosmological constant for now. This can be simplified further. Multiply through by $g_{\mu\nu}$ to get

$$R^{\mu}{}_{\mu} - \frac{1}{2} \delta^{\mu}{}_{\mu} R = 0. \tag{7.14}$$

Since $R^{\mu}{}_{\mu} \equiv R$ and $\delta^{\mu}{}_{\mu} = 4$, this implies $R = 0$ and hence, the vacuum Einstein equation is

$$R^{\mu\nu} = 0. \tag{7.15}$$

In Chapter 8, we will find one of its most important solutions.

Problems

7.1   Show that $R^{\mu\nu} = G^{\mu\nu} - \frac{1}{2} g^{\mu\nu} G$, where $G = G^{\mu}{}_{\mu}$.

7.2   An important property of the Riemann tensor is the second Bianchi identity

$$R^{\mu}{}_{\nu\rho\sigma;\tau} + R^{\mu}{}_{\nu\tau\rho;\sigma} + R^{\mu}{}_{\nu\sigma\tau;\rho} = 0.$$

Use this to demonstrate that the Einstein tensor is covariantly conserved, i.e.

$$\left(R^{\mu\nu} - \frac{1}{2} g^{\mu\nu} R \right)_{;\nu} = 0.$$

[Hint: think about the contraction which takes the Riemann tensor to the Ricci tensor. Don't forget the metric-compatibility condition and its implications discussed in Chapter 5.]

7.3* Show from equation (7.11) that the vacuum Einstein equation with a cosmological constant is

$$R^{\mu\nu} = g^{\mu\nu}\Lambda.$$

7.4* Show that the non-zero components of the Riemann tensor in the weak-field metric (7.10) are, to first order in Φ/c^2,

$$R_{0n0b} = \frac{1}{c^2}\Phi_{,n,b}, \quad R_{mnab} = \frac{1}{c^2}\left(-\delta_{mb}\Phi_{,n,a} + \delta_{nb}\Phi_{,m,a} + \delta_{ma}\Phi_{,n,b} - \delta_{na}\Phi_{,m,b}\right).$$

Hence, verify that the 00 component of the Einstein equation with zero cosmological constant is identical to the Poisson equation:

$$\nabla^2\Phi = 4\pi G\rho.$$

You may wish to refer back to Problem 6.6.

7.5 [Hard!] A conformally flat metric is one which can be written in the form $g_{\mu\nu} = \Phi^2(x)\eta_{\mu\nu}$, where $\eta_{\mu\nu}$ is the Minkowski metric, and Φ is a scalar function of the coordinates x. Find the equation for Φ implied by the Einstein equation. Compare it to Nordström's equation (4.12). If it is different, can you find a geometric equation involving curvature on the left-hand side and energy–momentum on the right-hand side which does give Nordström's equation?
[Hint: look at the trace of the Einstein equation.]

7.6 It has been proposed that one can relax the condition that the left-hand side of the Einstein equation contains only two derivatives of the metric. One can then look for another tensor, say $H^{\mu\nu}$, which contains four derivatives of the metric, and is covariantly conserved, so that the gravitational field equation would then take the form

$$G^{\mu\nu} + \Lambda g^{\mu\nu} + \beta H^{\mu\nu} = \frac{8\pi G}{c^4}T^{\mu\nu},$$

with β a constant. Show that the following expression:

$$H^{\mu\nu} = 2R^{;\mu;\nu} - 2g^{\mu\nu}R^{;\rho}_{\ ;\rho} - 2R^{\mu\nu}R + \frac{1}{2}g^{\mu\nu}R^2,$$

is a tensor satisfying the required conditions.

Chapter 8

The Schwarzschild Space–time

The Schwarzschild solution is the space–time outside a non-rotating spherical object · to get it, we compute Christoffel symbols, Riemann tensors, and Ricci tensors · we can use it in the Solar System

We are ready to undertake our first fully realistic calculation! This is to find the form of the space–time outside a static spherical object. It is known as the **Schwarzschild solution** or **Schwarzschild space–time**, after Karl Schwarzschild who discovered it in 1916, just a year after the theory first emerged. While it presumes that the source has zero rotation, in many circumstances, rotation is negligible and so the solution is widely applicable, for instance in describing the gravitational influence of the Sun on the planets.

We'll see that the solution itself is not too hard to obtain given the calculational technology we have assembled. On the other hand, the interpretation of the solution proves challenging, and it was perhaps only fifty years after the solution was first found that its full implications, particularly the concept of black holes, were properly understood.

We are interested only in what is going on outside the object providing the gravity, so the vacuum equations are what we need, $R^{\mu\nu} = 0$. The first stage is to find a suitable form of the metric. The guiding principles are the following:

- We don't expect the gravitational effects at a given location relative to the central object to vary with time, so there should exist coordinates in which the metric is time-independent.

- The object is spherically symmetric, so we can use spherical polar coordinates. Then the dependence of the metric on the angles θ and ϕ should follow from the spherical symmetry.

Introducing General Relativity, First Edition. Mark Hindmarsh and Andrew Liddle.
© 2022 John Wiley & Sons Ltd. Published 2022 by John Wiley & Sons Ltd.
Companion website: www.wiley.com/go/hindmarsh/introducingGR

- The gravitational forces should approach those predicted by Newtonian gravity at large radii.

- The metric should tend to the Minkowski form (in spherical polar coordinates) at infinity.

Together these imply we should consider the form:

$$ds^2 = -B(r)c^2 dt^2 + A(r) dr^2 + r^2 d\theta^2 + r^2 \sin^2\theta \, d\phi^2, \qquad (8.1)$$

where the two functions $A(r)$ and $B(r)$ should both tend to one at infinity.[1] This is the most general form of the metric consistent with our assumptions, and our task is to solve for the functions $A(r)$ and $B(r)$. Recall that for diagonal metrics, the components of the metric with raised indices equal one over those of the lowered index metric.

Our strategy mimics the toy two-dimensional example of the sphere that we worked through in Chapter 5:

1. Determine the Christoffel symbols from the metric.

2. Determine the Ricci tensor from the Christoffel symbols via the Riemann tensor.

3. Solve the Einstein equations which will result. Due to the spherical symmetry, there are only two independent equations, exactly the number required to find the two unknown functions.

However, now we are fully four-dimensional, so (ignoring symmetry properties) there are potentially 40 Christoffel symbols and 20 Riemann tensor components to worry about. Fortunately, we won't need them all, due to the symmetries of the metric.

In what follows, we will evaluate some of the quantities in detail, but give only results for the rest. The only way to really learn General Relativity is to go through such calculations in detail, so we strongly suggest you try to verify the stated results on your own.

8.1   Christoffel symbols

Exhaustive calculation will yield 13 non-zero Christoffel symbols (some of which follow immediately from symmetry under exchange of the lowered indices). The formula is

$$\Gamma^\mu{}_{\nu\sigma} = \frac{1}{2} g^{\mu\rho} \left(\partial_\sigma g_{\rho\nu} + \partial_\nu g_{\rho\sigma} - \partial_\rho g_{\nu\sigma} \right). \qquad (8.2)$$

---

[1] There are other ways to write these via coordinate redefinitions, which are ultimately equivalent.

The ones which only involve the angles θ and ϕ are the same as in two-dimensional spherical polar example from earlier; they are

$$\Gamma^\theta{}_{\phi\phi} = -\sin\theta\cos\theta \; ; \quad \Gamma^\phi{}_{\theta\phi} = \Gamma^\phi{}_{\phi\theta} = \frac{\cos\theta}{\sin\theta} . \tag{8.3}$$

As the metric is diagonal, we only get a contribution when the summed index ρ in equation (8.2) is the same as the raised index on the Christoffel symbol. Also, only derivatives of A and B with respect to r will be non-zero; we denote these by primes. Two worked examples are

$$\Gamma^r{}_{rr} = \frac{1}{2}g^{rr}\left(\partial_r g_{rr}\right) = \frac{A'}{2A} ; \tag{8.4}$$

$$\Gamma^t{}_{tr} = \Gamma^t{}_{rt} = \frac{1}{2}g^{tt}\left(\partial_r g_{tt}\right) = \frac{B'}{2B} . \tag{8.5}$$

The other non-zero ones are

$$\Gamma^r{}_{tt} = \frac{c^2 B'}{2A} ; \quad \Gamma^r{}_{\theta\theta} = -\frac{r}{A} ; \quad \Gamma^r{}_{\phi\phi} = -\frac{r\sin^2\theta}{A} ; \tag{8.6}$$

$$\Gamma^\theta{}_{r\theta} = \Gamma^\theta{}_{\theta r} = \Gamma^\phi{}_{r\phi} = \Gamma^\phi{}_{\phi r} = \frac{1}{r} . \tag{8.7}$$

8.2 Riemann tensor

As we are trying to determine two functions $A(r)$ and $B(r)$, there will be two independent equations. However, it is useful to determine three of the Einstein equations; those corresponding to $R^r{}_r$, $R^t{}_t$, and $R^\theta{}_\theta$. The definitions we need are

$$R^\mu{}_{\nu\alpha\beta} = \Gamma^\mu{}_{\beta\nu,\alpha} - \Gamma^\mu{}_{\alpha\nu,\beta} + \Gamma^\mu{}_{\alpha\rho}\Gamma^\rho{}_{\beta\nu} - \Gamma^\mu{}_{\beta\rho}\Gamma^\rho{}_{\alpha\nu} , \tag{8.8}$$

and

$$R_{\mu\nu} \equiv R^\alpha{}_{\mu\alpha\nu} . \tag{8.9}$$

Remember that the Riemann tensor is antisymmetric under exchange of the first two or last two indices, so

$$R^{tr}{}_{tr} = -R^{rt}{}_{tr} = -R^{tr}{}_{rt} = R^{rt}{}_{rt} . \tag{8.10}$$

We will need to compute six of these components. Again, only derivatives with respect to r will be non-zero. For example

$$
\begin{aligned}
R^t{}_{rtr} &= \Gamma^t{}_{rr,t} - \Gamma^t{}_{tr,r} + \Gamma^t{}_{t\rho}\,\Gamma^\rho{}_{rr} - \Gamma^t{}_{r\rho}\,\Gamma^\rho{}_{tr} \\
&= -\frac{1}{2}\frac{d}{dr}\frac{B'}{B} + \Gamma^t{}_{tr}\Gamma^r{}_{rr} - \left(\Gamma^t{}_{tr}\right)^2 \\
&= -\frac{1}{2}\left[\frac{B''}{B} - \left(\frac{B'}{B}\right)^2\right] + \frac{1}{4}\frac{A'}{A}\frac{B'}{B} - \frac{1}{4}\left(\frac{B'}{B}\right)^2 \\
&= -\frac{1}{2}\frac{B''}{B} + \frac{1}{4}\left(\frac{B'}{B}\right)^2 + \frac{1}{4}\frac{A'}{A}\frac{B'}{B}.
\end{aligned}
\tag{8.11}
$$

The others are

$$
R^t{}_{\theta t\theta} = -\frac{1}{2}\frac{r}{A}\frac{B'}{B};
\tag{8.12}
$$

$$
R^t{}_{\phi t\phi} = -\frac{1}{2}\frac{r\sin^2\theta}{A}\frac{B'}{B};
\tag{8.13}
$$

$$
R^r{}_{\theta r\theta} = \frac{1}{2}\frac{r}{A}\frac{A'}{A};
\tag{8.14}
$$

$$
R^r{}_{\phi r\phi} = \frac{1}{2}\frac{r\sin^2\theta}{A}\frac{A'}{A};
\tag{8.15}
$$

$$
R^\theta{}_{\phi\theta\phi} = \sin^2\theta\left(1 - \frac{1}{A}\right).
\tag{8.16}
$$

8.3   Ricci tensor

Now, we sum Riemann tensor components with the same first and third indices to get the components of the Ricci tensor. We need to use the symmetries of the Riemann tensor, but we must be careful to raise and lower indices using the metric. For example, $R^{rt}{}_{trt} = g_{tt}R^{rt}{}_{rt} = g_{tt}R^{tr}{}_{tr} = g_{tt}g^{rr}R^t{}_{rtr} = -(c^2B/A)R^t{}_{rtr}$.

Let's go for it!

$$
\begin{aligned}
R_{tt} &= R^r{}_{trt} + R^\theta{}_{t\theta t} + R^\phi{}_{t\phi t} \\
&= -\frac{c^2B}{A}\left[-\frac{1}{2}\frac{B''}{B} + \frac{1}{4}\left(\frac{B'}{B}\right)^2 + \frac{1}{4}\frac{A'}{A}\frac{B'}{B}\right] - \frac{c^2B}{r^2}\left[-\frac{1}{2}\frac{r}{A}\frac{B'}{B}\right] \\
&\quad - \frac{c^2B}{r^2\sin^2\theta}\left[-\frac{1}{2}\frac{r\sin^2\theta}{A}\frac{B'}{B}\right] \\
&= -\frac{c^2B}{A}\left[-\frac{1}{2}\frac{B''}{B} + \frac{1}{4}\left(\frac{B'}{B}\right)^2 + \frac{1}{4}\frac{A'}{A}\frac{B'}{B} - \frac{1}{r}\frac{B'}{B}\right].
\end{aligned}
\tag{8.17}
$$

The others we need are

$$R_{rr} = -\frac{1}{2}\frac{B''}{B} + \frac{1}{4}\left(\frac{B'}{B}\right)^2 + \frac{1}{4}\frac{A'}{A}\frac{B'}{B} + \frac{1}{r}\frac{A'}{A} ; \tag{8.18}$$

$$R_{\theta\theta} = -\frac{r}{A}\left[\frac{1}{2}\frac{B'}{B} - \frac{1}{2}\frac{A'}{A} - \frac{1}{r}(A-1)\right] . \tag{8.19}$$

8.4 The Schwarzschild solution

Next, we have to solve the equations we get from equating the Ricci tensor components to zero. The reason for computing both R_{rr} and R_{tt} is that they combine nicely to give a simple equation. Take $AR_{tt}/c^2B + R_{rr}$ and set it to zero to get

$$\frac{B'}{B} = -\frac{A'}{A} . \tag{8.20}$$

Put this in the right-hand side of equation (8.19) and set the result to zero, to get

$$\frac{A'}{A} = -\frac{1}{r}(A-1) . \tag{8.21}$$

Taking advantage of separability this can be solved to give

$$\ln A - \ln(A-1) = \ln r - \ln a , \tag{8.22}$$

where a is the constant of integration, which rearranges to

$$A(r) = \left(1 - \frac{a}{r}\right)^{-1} . \tag{8.23}$$

Meanwhile, equation (8.20) implies that $A = b/B$ where b is another constant. But to get the flat-space solution, we need B to tend to a constant as $r \to \infty$; moreover, a residual freedom to rescale the time coordinate allows that constant to be chosen to be unity. With this choice, $b = 1$, and we have

$$B(r) = 1 - \frac{a}{r} . \tag{8.24}$$

Finally, we need to interpret the constant a. We showed in Chapter 7 that in a weak gravitational field, the space–time interval is approximately given by equation (7.10) and hence, $g_{tt} = -(1 + 2\Phi/c^2)$, with Φ the Newtonian gravitational potential. In the Schwarzschild solution, $g_{tt} = -B(r)$. The Correspondence Principle demands that they are the same in the limit $r \to \infty$. Since the Newtonian potential around a spherical body is $\Phi = -GM/r$, we have

$$a = \frac{2GM}{c^2}.\tag{8.25}$$

So the constant a contains information about the mass of the body producing the gravitational field.

We have arrived at the Schwarzschild metric! It is

$$ds^2 = -c^2 \left(1 - \frac{2GM}{rc^2}\right) dt^2 + \left(1 - \frac{2GM}{rc^2}\right)^{-1} dr^2 + r^2 d\theta^2 + r^2 \sin^2\theta \, d\phi^2.$$
$$\tag{8.26}$$

The constant a has dimensions of length and has important significance which will be discussed in Chapter 11. For now, we will just note that it is called the Schwarzschild radius, and reserve the symbol r_S for it, so that

$$r_S \equiv \frac{2GM}{c^2}.\tag{8.27}$$

8.5   The Jebsen–Birkhoff theorem

In this chapter, we found the solution outside a non-rotating spherical body under the assumption that the solution is static and tends to Minkowski space at infinity. This is the Schwarzschild solution, which is a vacuum solution.

In fact, it can be proven, rather than assumed, that the exterior solution outside a non-rotating spherical body has to be static and asymptotically flat as a consequence of Einstein's equation. This is the content of theorems proved first by the Norwegian mathematician Jørg Tofte Jebsen in 1921 and then independently by the American George Birkhoff in 1923, who was unaware of Jebsen's paper.[2]

The important consequence of the theorem is that the Schwarzschild solution is the *only* static spherically symmetric vacuum solution, and hence, it must apply outside any spherically symmetric matter distribution, even if the matter distribution sourcing the gravitational field is non-static. The Schwarzschild solution therefore holds outside objects like stars, whose rotation is slow enough to be ignored, and even applies outside a spherical body which is pulsating or collapsing radially inwards. In particular, the Schwarzschild solution applies of course to the Sun, and it can be used to describe motions within the Solar System.

In Problem 8.6, you can work through the steps required to derive the solution inside a star as well as outside. We will study the motion of test particles in the Schwarzschild space–time in Chapter 9.

---

[2]Knowledge of Jebsen's work gradually disappeared in the 1920s, perhaps because of his early death in 1922, and so you will often see it called just Birkhoff's theorem.

Problems

8.1* The following calculations refer to the Schwarzschild geometry:

a) Starting from the Christoffel symbols given in Section 8.1, compute the Riemann tensor component $R^{\theta\phi}_{\;\;\theta\phi}$, which is the sectional curvature of surfaces of constant r and t (see Section 5.8). Compare it to the curvature of a two-sphere with radius r in Euclidean space.

b) Compute the Riemann tensor component $R^{tr}_{\;\;tr}$, which is the sectional curvature of the rt surfaces. Evaluate it at the Schwarzschild radius and at $r = 0$, and comment.

8.2* Considering the Schwarzschild metric, define a new radial coordinate $\tilde{r}$ in place of r by

$$r = \tilde{r}\left(1 + \frac{GM}{2c^2\tilde{r}}\right)^2 .$$

Write the Schwarzschild metric in the simplest form possible using this new coordinate. This is known as the isotropic form, and the coordinates as isotropic coordinates. Hence demonstrate that, in these coordinates, the metric at large distances takes the form derived in the weak-field limit in Section 7.3,

$$ds^2 = -\left(1 + \frac{2\Phi}{c^2}\right)dt^2 + \left(1 - \frac{2\Phi}{c^2}\right)\left(d\tilde{r}^2 + \tilde{r}^2 d\theta^2 + \tilde{r}^2 \sin^2\theta\, d\phi^2\right),$$

with Φ the Newtonian potential at distance $\tilde{r}$ from a body of mass M.

8.3* Use the vacuum Einstein equation with cosmological constant derived in Problem 7.3 to show that the spherically symmetric solution has

$$B(r) = A^{-1}(r) = 1 - \frac{r_s}{r} - \frac{\Lambda r^2}{3} .$$

This is called the Schwarzschild–de Sitter solution. What does this imply about the acceleration of a test body at radii $(r_s/\Lambda)^{1/3} \ll r \ll \Lambda^{-1/2}$? (You may assume that $r \gg r_s$.)

8.4 The metric around a rotating object with axial symmetry is known as the

Kerr metric, whose space–time interval can be written as

$$ds^2 = -c^2 \left(1 - \frac{r_S r}{\rho^2}\right) dt^2 - 2\frac{r r_S a \sin^2 \theta}{\rho^2} dt d\phi + \left(\frac{\rho^2}{r^2 + a^2 - r_S r}\right) dr^2$$
$$+ \rho^2 d\theta^2 + \left(r^2 + a^2 + \frac{r r_S a^2 \sin^2 \theta}{\rho^2}\right) \sin^2 \theta \, d\phi^2,$$

where r_S and a are constants, and $\rho^2 = r^2 + a^2 \cos^2 \theta$. The constant a is the angular momentum per unit mass, divided by the speed of light c.

a) Show that this metric reduces to the Schwarzschild one in the limit $a \to 0$.

b) Find the metric when $r_S \to 0$. Show that this is Minkowski space in oblate spheroidal coordinates,

$$x = (r^2 + a^2)^{1/2} \sin \theta \cos \phi, \quad y = (r^2 + a^2)^{1/2} \sin \theta \sin \phi, \quad z = r \cos \theta.$$

c) What extra non-zero Christoffel symbols are there for this metric, compared to the Schwarzschild metric? You don't need to calculate their values. In which Christoffel symbols does the inverse metric component $g^{t\phi}$ appear?

8.5* Consider the general spherically symmetric and static metric (8.1), and write $A(r) = e^{2\psi(r)}$, $B(r) = e^{2\phi(r)}$.

a) Write out the non-zero components of the Christoffel symbols given in Section 8.1 in terms of ϕ and ψ.

b) Find the Ricci tensor in terms of ϕ and ψ, and show that the non-zero components of the Einstein tensor are

$$G^t_{\ t} = -\frac{1}{r^2} \left[2r\psi' e^{-2\psi} + \left(1 - e^{-2\psi}\right)\right];$$

$$G^r_{\ r} = \frac{1}{r^2} \left[2r\phi' e^{-2\psi} - \left(1 - e^{-2\psi}\right)\right];$$

$$G^\theta_{\ \theta} = G^\phi_{\ \phi} = e^{-2\psi} \left[\phi'' + (\phi')^2 - \phi'\psi' + \frac{1}{r}(\phi' - \psi')\right].$$

These results will be useful in Problem 8.6.

8.6 [Hard!] In this problem, you can derive the *Tolman–Oppenheimer–Volkov* (TOV) equation, whose solution describes the interiors of stars, modelled as a gravitating ball of perfect fluid, according to General Relativity. In this question, we use the metric in the same form as Problem 8.5.

a) Show that for a stationary perfect fluid with 4-velocity u^μ in the general spherically symmetric metric, $u^0 = ce^{-\phi}$, and find the components of the energy–momentum tensor in mixed form $T^\mu{}_\nu$.

b) Explain why the only non-zero term in the energy–momentum conservation equation is $T^{rr}{}_{;r}$, and use the results from Problem 8.5 to show that it implies

$$\frac{dp}{dr} + \phi'(\rho c^2 + p) = 0,$$

where p and ρ are the pressure and density of the fluid.

c) Use the expression for the Einstein tensor in Problem 8.5 to show that the tt and rr components of the Einstein equation can be rearranged to give

$$\frac{dm}{dr} = \frac{1}{2}\kappa r^2 \rho c^2, \quad \frac{d\phi}{dr} = \frac{2m + \kappa r^3 p}{2r(r - 2m)},$$

where $m = r(1 - e^{-2\psi})/2$, and $\kappa = 8\pi G/c^4$.

d) Outside the fluid, where ρ and p vanish, show that $\psi, -\phi \to GM/rc^2$ at large r, where $M = 4\pi \int dr\, r^2 \rho$. This explicitly demonstrates that the constant in the Schwarzschild solution does indeed match the mass of the body in the centre, and that the mass is the integral of the energy density.

e) Combine the equations you have found to derive the TOV equation:

$$\frac{dp}{dr} = -\frac{(\rho c^2 + p)(\kappa p r^3 + 2m)}{2r(r - 2m)}.$$

When the equation of state of the fluid $p = p(\rho)$ is known, the pair of equations for m' and p' can be solved.

Chapter 9

Geodesics and Orbits

*If there are no external forces, particles move on geodesics ·
curvature causes geodesics to converge or diverge · orbits are the
geodesics around a massive object*

9.1 Geodesics

Geodesics generalise the idea of straight lines to curved space.

Let's think about what we mean when we say a line is straight in ordinary Euclidean space. One way of defining a straight line is the shortest distance between two points. Another is a curve with constant direction. Both ways can be generalised to Riemannian space–time. We will take the second definition as our starting point.

Let us consider a curve in Euclidean space with Cartesian coordinates $x^i(\lambda)$, where λ is a parameter that measures the location along the curve. The tangent vector has Cartesian components

$$t^i = \frac{dx^i}{d\lambda}. \tag{9.1}$$

If we move along the curve, the direction of motion is along the unit vector $\hat{t}^i = t^i/|\mathbf{t}|$, where $|\mathbf{t}| = \sqrt{\mathbf{t}^2}$. The rate of change of the direction along the curve is

$$\frac{d\hat{t}^i}{d\lambda} = \frac{1}{|\mathbf{t}|}\left[\frac{dt^i}{d\lambda} - \frac{1}{2}\frac{1}{\mathbf{t}^2}\left(\frac{d(\mathbf{t}^2)}{d\lambda}\right)t^i\right]. \tag{9.2}$$

Introducing General Relativity, First Edition. Mark Hindmarsh and Andrew Liddle.
© 2022 John Wiley & Sons Ltd. Published 2022 by John Wiley & Sons Ltd.
Companion website: www.wiley.com/go/hindmarsh/introducingGR

A straight line therefore satisfies

$$\frac{dt^i}{d\lambda} = \alpha(\lambda)t^i, \tag{9.3}$$

where $\alpha(\lambda)$ is the rate of change of the length of the vector. The direction of this curve does not change, but the length of its tangent vector does.

One can always choose the parameter λ such that the length of the tangent vector is constant, i.e. $d(\mathbf{t}^2)/d\lambda = 0$, so that a straight line satisfies (see Problem 9.1)

$$\frac{dt^i}{d\lambda} = 0. \tag{9.4}$$

In this case, the parameter λ is said to be **affine**. Any parameter related linearly to λ, i.e. $\lambda' = a\lambda + b$ with a, b constants, is also affine since $d\lambda'/d\lambda$ is a constant.

One can think of Newton's First Law as the statement that, in the absence of external forces, the trajectory of a particle $\mathbf{x}(t)$ is a straight line, and that time is an affine parameter.

The notion of a straight line also exists in Minkowski space, where a curve has coordinates $x^\mu(\lambda)$. Points on the curve move through time as well as space, and the tangent vector $t^\mu \equiv dx^\mu/d\lambda$ is a 4-vector. The scalar product of the tangent vector with itself, $t^2 \equiv t^\mu t_\mu$, can be positive, zero, or negative.

If the curve is a straight line (from now on let's start using the word geodesic) t^2 is a constant in an affine parametrisation, and so must always be positive, zero, or negative. Hence, we can divide geodesics in Minkowski space into three categories, just like any 4-vector: spacelike ($t^2 > 0$), null ($t^2 = 0$), and timelike ($t^2 < 0$). Because of the Lorentz invariance of the scalar product, all observers agree on which category a geodesic falls into.

Let us now think about geodesics in curved space and space–time. We are familiar with the idea that on the surface of the Earth, which is approximately a two-sphere, if one starts moving in a constant direction one will eventually trace out a great circle. In order to describe the idea mathematically, we would like to generalise the definition, equation (9.4), but we first have to overcome a problem stemming from the definition of the derivative of the tangent vector.

In Minkowski space, and with Cartesian coordinates, we have

$$\frac{dt^\mu}{d\lambda} = \lim_{\delta\lambda \to 0} \frac{t^\mu(\lambda + \delta\lambda) - t^\mu(\lambda)}{\delta\lambda}, \tag{9.5}$$

and there is no difficulty with the right-hand side. However, as we saw in Chapter 5, we cannot directly subtract vectors at two different points in a curved space; we must parallel transport one of them to the location of the other (see Figure 9.1).

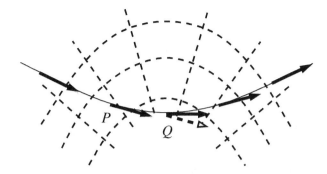

Figure 9.1 In order to define the derivative of the tangent vector at point Q, the tangent vector at neighbouring point P must be parallel transported to Q (indicated by the dashed arrow) before the difference is taken.

Under parallel transport along the curve from point P to point Q by an amount $\delta\lambda$, the components of the contravariant vector $t^\mu(\lambda)$ change by

$$\delta_\parallel t^\mu(\lambda) = -\Gamma^\mu_{\nu\sigma} t^\nu \delta x^\sigma , \tag{9.6}$$

where $\delta x^\rho = t^\rho \delta\lambda$. The covariant derivative of the tangent vector in the direction $dx^\mu = t^\mu d\lambda$ is

$$\begin{aligned}
\frac{D_t t^\mu}{d\lambda} &= \lim_{\delta\lambda\to 0} \frac{t^\mu(\lambda+\delta\lambda) - \left[t^\mu(\lambda) + \delta_\parallel t^\mu(\lambda)\right]}{\delta\lambda} ; \\
&= \frac{dt^\mu}{d\lambda} + \Gamma^\mu_{\nu\sigma} t^\nu t^\sigma ,
\end{aligned} \tag{9.7}$$

where we use the notation D_t, with the subscript indicating the vector along which we project the covariant derivative. The generalisation of the notion of a geodesic to a curved space is then the following:

> A **geodesic** is a curve whose tangent vector remains constant if parallel transported along the curve.

That is, a geodesic in a space–time with Christoffel symbols $\Gamma^\mu_{\nu\sigma}$ is a curve $x^\mu(\lambda)$ with a tangent vector $t^\mu(\lambda)$ obeying

$$\frac{dt^\mu}{d\lambda} + \Gamma^\mu_{\nu\sigma} t^\nu t^\sigma = 0, \tag{9.8}$$

or equivalently

$$\frac{d^2x^\mu}{d\lambda^2} + \Gamma^\mu_{\nu\sigma}\frac{dx^\nu}{d\lambda}\frac{dx^\sigma}{d\lambda} = 0. \qquad (9.9)$$

These are two versions of what we call the **geodesic equation**.

In General Relativity, the generalisation of Newton's First Law is:

In the absence of external forces, particles move on geodesics.

Geodesics tell us how objects move in the presence of curved space–time, but do not tell us how the curvature of space–time is generated in the first place.

We saw that geodesics with an affine parameter have a constant magnitude. Hence, the division into spacelike, null, and timelike carries over into General Relativity as well. Indeed, with a simple rescaling, we can choose the affine parameter for a timelike tangent vector such that $t^2 = -c^2$. With this choice, the affine parameter is the proper time, and the tangent vector is the 4-velocity $u^\mu = dx^\mu/d\tau$. Ordinary massive particles move on timelike geodesics with $u^2 = -c^2$, while light and other massless particles move on null geodesics, $u^2 = 0$. No physical object moves on a spacelike geodesic, although sometimes the word *tachyon* is used to describe an unphysical particle with spacelike 4-velocity.

We can summarise the conditions on the 4-velocities of physical particles as follows:

$$\frac{dx^\mu}{d\tau}g_{\mu\nu}(x(\tau))\frac{dx^\nu}{d\tau} = \begin{cases} -c^2 & \text{massive particle} \\ 0 & \text{massless particle} \end{cases}. \qquad (9.10)$$

These are quite non-trivial equations in a general space–time. We will see how they can be solved in the special case of the Schwarzschild metric later in this chapter.

9.2   Non-relativistic limit of geodesic motion

Let us suppose that the metric is static, or at least changing slowly, and that the particle speed v is much less than the speed of light c. The proper time τ is then approximately the coordinate time t, and the geodesic equation for the space coordinates simplifies to

$$\frac{d^2x^i}{dt^2} + c^2\Gamma^i_{00} + O(vc) = 0. \qquad (9.11)$$

In a static metric, the relevant Christoffel symbols reduce to

$$\Gamma^i_{00} = g^{ij}\partial_j g_{00}. \qquad (9.12)$$

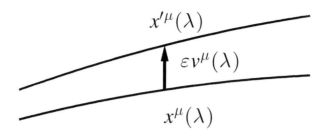

Figure 9.2   Geodesic deviation: in a curved space–time, nearby geodesics may move apart or come together. Consider points $x^\mu(\lambda)$ and $x'^\mu(\lambda)$ on neighbouring geodesics $x^\mu(\lambda)$, separated by a small displacement εv^μ. The vector v^μ evolves according to the geodesic equation (9.19), which directly involves the Riemann curvature.

We have already argued in Section 4.6, on the basis of gravitational time dilation, that the 00 component of the metric contains the Newtonian gravitational potential Φ in the form $g_{00} \simeq -(1+2\Phi/c^2)$, in weak gravitational fields where coordinates can be chosen so that the metric is close to the Minkowski metric. Here is more evidence for that identification, as the resulting equation for the particle motion is

$$\frac{d^2 x^i}{dt^2} = -\partial^i \Phi. \tag{9.13}$$

This is just the Newtonian equation for the acceleration of a particle in a gravitational potential Φ.

Hence, the geodesic equation, which is the General Relativistic version of Newton's First Law, automatically includes the gravitational force. It also incorporates the Weak Equivalence Principle: all bodies are accelerated equally as they move through space–time.

9.3   Geodesic deviation

Another important equation is how nearby geodesics behave. In flat space–time, two parallel lines remain forever parallel, but in the presence of curvature this is no longer true.

Suppose we have two neighbouring geodesics $x^\mu(\lambda)$ and $x'^\mu(\lambda)$ (see Figure 9.2). We define a vector v^μ, such that

$$x'^\mu(\lambda) = x^\mu(\lambda) + \varepsilon v^\mu(\lambda), \tag{9.14}$$

where ε is small. The geodesics satisfy the equations

$$\frac{D_{t'}t'^{\mu}}{d\lambda} = \frac{d^2x'^{\mu}}{d\lambda^2} + \Gamma^{\mu}_{\nu\sigma}(x')\frac{dx'^{\nu}}{d\lambda}\frac{dx'^{\sigma}}{d\lambda} = 0; \qquad (9.15)$$

$$\frac{D_t t^{\mu}}{d\lambda} = \frac{d^2x^{\mu}}{d\lambda^2} + \Gamma^{\mu}_{\nu\sigma}(x)\frac{dx^{\nu}}{d\lambda}\frac{dx^{\sigma}}{d\lambda} = 0, \qquad (9.16)$$

where $t = dx^{\mu}/d\lambda$ is the tangent vector to the geodesic $x^{\mu}(\lambda)$, and similar expression holds for t'^{μ}.

Our strategy for determining how these geodesics drift apart or together will be to expand the first equation to first order in the parameter ε, and attempt to discover a covariant equation for the separation vector v^{μ}.

Expanding equation (9.15) in powers of ε, we find

$$\frac{D_{t'}t'^{\mu}}{d\lambda} = \frac{D_t t^{\mu}}{d\lambda} + \varepsilon\left(\frac{d^2v^{\mu}}{d\lambda^2} + 2\Gamma^{\mu}_{\nu\rho}\frac{dv^{\nu}}{d\lambda}t^{\rho} + \Gamma^{\mu}_{\nu\rho,\sigma}t^{\nu}t^{\rho}v^{\sigma}\right) + O(\varepsilon^2), \quad (9.17)$$

where we have used the symmetry of the Christoffel symbols. Hence, subtracting equation (9.16), dividing by ε, and taking the limit $\varepsilon \to 0$, we have

$$\frac{d^2v^{\mu}}{d\lambda^2} + 2\Gamma^{\mu}_{\nu\rho}\frac{dv^{\nu}}{d\lambda}t^{\rho} + \Gamma^{\mu}_{\nu\rho,\sigma}t^{\nu}t^{\rho}v^{\sigma} = 0. \qquad (9.18)$$

This must be a covariant equation in disguise. The derivative of the Christoffel symbols means that there must be a term involving the Riemann tensor, and the second derivative of the vector with respect to the affine parameter along the curve $x^{\mu}(\lambda)$ could only come from a term $D_t^2v^{\mu}/d\lambda$.

The equation for the separation vector between neighbouring geodesics v^{μ} is therefore

$$\frac{D_t^2 v^{\mu}}{d\lambda^2} = R^{\mu}_{\nu\rho\sigma}\frac{dx^{\nu}}{d\lambda}\frac{dx^{\sigma}}{d\lambda}v^{\rho}, \qquad (9.19)$$

where we recall that $t^{\mu} = dx^{\mu}/d\lambda$ is the tangent vector to the reference geodesic. This is known as the **geodesic deviation equation**. We leave the verification that equations (9.18) and (9.19) are the same equation as an exercise, Problem 9.2.

Equation (9.19) is the mathematical expression of the physical effect we already noted in Section 4.5; that objects freely falling in a gravitational field (i.e. following geodesics) do not maintain their relative distance in an inertial frame. It is indeed the curvature of space–time which determines how two neighbouring geodesics drift apart or together.

In Problem 9.3, you can use the results for the Riemann tensor outside a body of mass M obtained in Section 8.2 to verify that two non-relativistic test particles freely falling towards the centre of a massive body accelerate towards one another with the Newtonian value (4.20). It will help to first study the motion of freely falling particles, that is the orbits, in the Schwarzschild metric.

9.4 Newtonian theory of orbits

Before studying the orbits of bodies in the Schwarzschild metric, we will first review the Newtonian theory of orbits, as in many circumstances, the General Relativistic version can be viewed as a small perturbation to the Newtonian result.

We first write down the energy of a particle of mass m moving in the Newtonian potential of a body of mass M. We will assume that $M \gg m$ so that the large body can be taken to be stationary, and use polar coordinates (r, θ, ϕ). Conservation of angular momentum assures us that a particle in a central potential always moves in a single plane, which we can take to be the equatorial plane $\theta = \pi/2$.

The total energy is given by

$$E = \frac{1}{2}m\left(\frac{dr}{dt}\right)^2 + \frac{1}{2}mr^2\left(\frac{d\phi}{dt}\right)^2 + m\Phi(r), \tag{9.20}$$

where $\Phi(r) = -GM/r$ is the gravitational potential. The energy is conserved, as is the angular momentum. The only non-zero component is in the direction perpendicular to the orbit plane,

$$L_\phi = mr^2\frac{d\phi}{dt}. \tag{9.21}$$

Hence, we can write

$$e = \frac{1}{2}\left(\frac{dr}{dt}\right)^2 + V_{\text{eff}}(r), \tag{9.22}$$

where $e = E/m$ and $\ell = L_\phi/m$ are the specific energy and angular momentum (i.e. the energy and angular momentum per unit mass), and V_{eff} is an 'effective' potential, given by

$$V_{\text{eff}}(r) = -\frac{GM}{r} + \frac{\ell^2}{2r^2}. \tag{9.23}$$

We can now view the system as being a particle with specific energy e moving in one (radial) dimension, in a potential V_{eff}. This potential is sketched in Figure 9.3. Differentiation shows that it has a minimum $V_{\text{min}} = -G^2M^2/2\ell^2$ at $r_{\text{min}} = \ell^2/GM$.

One can see that if the specific energy e is less than zero, but greater than V_{min},

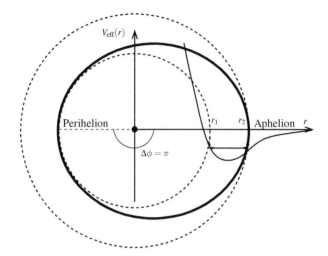

Figure 9.3 The orbit of a particle in a central potential, according to Newtonian theory. Superimposed is the effective potential $V_{eff}(r)$ governing the radial motion of the particle, which moves between maximum and minimum distances r_2 and r_1, as the polar angle ϕ changes by exactly π. The angle therefore changes by 2π when the particle returns to its original radius, and the orbit closes. For orbits around the Sun, the points of furthest and closest approach are called aphelion and perihelion.

the particle moves between minimum and maximum radii r_1 and r_2, defined by the points at which $dr/dt = \sqrt{2(e - V_{eff})} = 0$. If the orbit in question is around the Sun, we call the points on the orbit with minimum and maximum radii the perihelion and aphelion; around any other star they are called periastron and apastron, while the general terms for orbits where the bodies are not specified are periapsis and apoapsis.

Considering orbits around the Sun, we can calculate the change in the angular coordinate ϕ between perihelion and aphelion by noting that

$$\frac{d\phi}{dr} = \frac{d\phi/dt}{dr/dt} = \frac{1}{r^2} \frac{\ell}{\sqrt{2(e - V_{eff}(r))}}. \tag{9.24}$$

The r integral can be performed with limits r_1 and r_2, to show that the change in polar angle between aphelion and perihelion is[1]

$$\Delta\phi = \pi, \tag{9.25}$$

---

[1] See Advanced Topic A2 for a fuller explanation.

regardless of the orbital parameters. In other words, the particle (or planet) moves through exactly half a revolution as it travels from perihelion to aphelion. By symmetry, it will move though another half-revolution as it travels back to aphelion. Orbits in a Newtonian potential are therefore closed, and it is not too difficult to show that the shape of the orbit is an ellipse, with the central body at one of the foci (see Figure 9.3). These orbits obey Kepler's three laws — concerning shape, period, and orbital speed — and hence are also known as Keplerian orbits.

It is useful to recall some nomenclature for ellipses, which are often described in terms of the semi-major axis a, the semi-minor axis b, and the eccentricity ε. The relations between these quantities and the minimum and maximum radii r_1 and r_2 are

$$a = \frac{r_1 + r_2}{2}, \quad b = \sqrt{r_1 r_2}, \quad \varepsilon = \frac{r_2 - r_1}{r_2 + r_1}. \tag{9.26}$$

In terms of these parameters, the equation of the ellipse is given by

$$r = \frac{a(1 - \varepsilon^2)}{1 + \varepsilon \cos \phi}, \tag{9.27}$$

where the angle ϕ is conventionally measured from the point of closest approach.

These results can be extended to the case where $e > 0$ and the orbit is unbound, and there is only one turning point, at $r = r_1$. Formally, r_2 and a become negative, and the eccentricity greater than unity. Under these conditions, equation (9.27) remains valid and describes a hyperbola.

9.5 Orbits in the Schwarzschild space–time

9.5.1 Massive particles

In this section, we give a brief treatment of orbits of massive particles in the Schwarzschild space–time, with a detailed derivation of the dynamical equations reserved for Advanced Topic A1. In General Relativity, particles move on geodesics, whose tangent vector is the 4-velocity u^μ, and whose magnitude is constant. In the Schwarzschild background, the resulting equation $u^2 = -c^2$ can be rearranged to look like the Newtonian energy equation (9.22)

$$\mathcal{E} = \frac{1}{2} \left(\frac{dr}{d\tau} \right)^2 + V_{\text{eff}}^{\text{GR}}(r), \tag{9.28}$$

where τ is the proper time, and the general relativistic version of the effective potential is

$$V_{\text{eff}}^{\text{GR}}(r) = -\frac{GM}{r} + \frac{\ell^2}{2r^2} - \frac{GM\ell^2}{c^2 r^3}.$$ (9.29)

In the expression for the potential, $\mathscr{E}$ and ℓ are constants, with ℓ the angular momentum per unit rest mass, or

$$\ell = r^2 \frac{d\phi}{d\tau}.$$ (9.30)

Hence, ℓ/r is the azimuthal component of the 4-velocity. When velocities are small compared with c, the factor $\ell^2/r^2 c^2$ is small, and proper time and coordinate time approach each other. Hence, equation (9.28) reduces to

$$\mathscr{E} \simeq \frac{1}{2}\left(\frac{dr}{dt}\right)^2 + \frac{\ell^2}{2r^2} - \frac{GM}{r},$$ (9.31)

which has the same form as the Newtonian orbit equation (9.22), as the Correspondence Principle demands. While the constant $\mathscr{E}$ is not strictly a specific energy, it still has units of velocity squared, and reduces to the Newtonian specific energy when $\mathscr{E} \ll c^2$.

Once again, we can plot the effective potential and qualitatively determine the orbit types from the value of $\mathscr{E}$ with respect to the effective potential.

It is usually convenient to write expressions in terms of the Schwarzschild radius $r_S \equiv 2GM/c^2$, because G and M always occur together in this combination. For ordinary stars and planets, this is a purely notional radius, as it is well inside the body and the particle can never reach it. One can also show that in near-circular orbits, as relevant for planets in our Solar System, $\ell \gg r_S c$.

In this case, the effective potential in General Relativity resembles that in Figure 9.4. There are two extrema. At large r, it has a qualitatively similar shape to the Newtonian one: it has a local minimum and tends to zero at large radius. At small radii, the potential becomes large and positive, which can be interpreted as the effect of the angular momentum keeping the particle away from the centre.

However, as $r \to 0$, the relativistic correction to the potential becomes more important and eventually dominates. Instead of tending to positive infinity, the potential reaches a local maximum, and tends to negative infinity at the centre. Again, these features would be well inside an ordinary astronomical object. Only if the object sourcing the gravitational field is a black hole, as described in Chapter 11, are these parts of the effective potential relevant.

Denoting the values of the effective potential at its turning points by V_{max} and V_{min}, the orbit types for ordinary astronomical bodies are as follows.

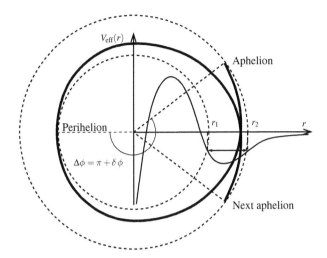

Figure 9.4 Orbit of a particle in a central potential with $\ell \gg r_S c$, according to General Relativity. Superimposed is the effective potential $V_{\text{eff}}(r)$ governing the radial motion of a particle, which moves between maximum and minimum distances r_2 and r_1, while in this case the polar angle ϕ changes by more than π. The angle therefore increases by more than 2π when returning to its original radius, and the orbit does not close. The angles of aphelion and perihelion increase or *precess* with each orbit by an amount $2\delta\phi$.

- $\mathcal{E} = V_{\text{min}}$: A stable circular orbit at $r \simeq 2(\ell/r_S c)^2 r_S$.

- $V_{\text{min}} < \mathcal{E} < 0$: Solutions whose radius varies between a maximum and a minimum. These orbital solutions are the general relativistic analogue of Newtonian elliptical orbits.

- $0 \leq \mathcal{E} < V_{\text{max}}$: The particle comes in from infinity and (if it does not hit the central body) rebounds back to infinity, analogous to a hyperbolic Newtonian orbit. In General Relativity, if $\mathcal{E}$ is close to V_{max}, the particle can rotate around the central body many times before being flung out to infinity again.

The class of obvious physical relevance is the analogue of Newtonian elliptical orbits, which applies to Solar System planetary orbits. While Newtonian orbits close exactly on themselves, Schwarzschild orbits do not, giving rise to the perihelion shift that has been measured for the four inner planets of the Solar System, most accurately for Mercury which was the only measured shift at the time Einstein was developing General Relativity. Further development of the maths given above allows this shift to be calculated, which we shall do in Chapter 10.

9.5.2   Photon orbits

The analysis of the motion of massless particles such as photons has the complication that there is no Newtonian limit for comparison: if we try to proceed as in Section 9.4, we immediately encounter the problem of writing down a Newtonian expression for the kinetic energy of a massless particle. As the mass drops out of the equations of motion, we could just overlook the problem. With General Relativity, we are on much more secure footing, as we understand that massless particles move along null geodesics. In Advanced Topic A1, we will see that the radial coordinate is a function of an affine parameter λ, which takes on the role of measuring the location along the trajectory previously undertaken by proper time τ. In a similar way to equation (9.28), the radial coordinate of the photon satisfies

$$\frac{1}{b^2} = \left(\frac{dr}{d\lambda}\right)^2 + W_{\text{eff}}(r) , \tag{9.32}$$

where

$$W_{\text{eff}}(r) = \frac{1}{r^2}\left(1 - \frac{r_S}{r}\right) , \tag{9.33}$$

and $b = \ell/e$ is the ratio of the angular momentum and the energy of the photon, with an appropriate factor of c to give it the dimensions of length. With this choice of affine parameter, the equation for the angular coordinate is

$$r^2\frac{d\phi}{d\lambda} = 1 . \tag{9.34}$$

The constant b has the interpretation of an impact parameter: the distance between the geodesic and a parallel line passing through the central mass.[2] That is, as seen in Figure 9.5, it is the distance that would have been the one of closest approach had the photon trajectory not been curved.

Referring to Figure 9.5, one sees that qualitatively different behaviour is found according to the relative size of the maximum of the effective potential W_{max} and $1/b^2$. For ordinary astronomical bodies, the impact parameter satisfies $b^2 > 1/W_{\text{max}}$. In this case, the radial coordinate decreases to a minimum r_1 given by $W_{\text{eff}}(r_1) = 1/b^2$ and then increases again. This corresponds to the photon approaching from infinity, suffering a bend in its trajectory, and then escaping back to infinity on a modified trajectory (unless the object producing the gravitational field itself has radius bigger than r_1, in which case the photon will hit it). We will calculate the deflection angle in Chapter 10.

---

[2]There is no difficulty about defining the notion of parallel at infinity, where the space–time metric approaches the Minkowski one.

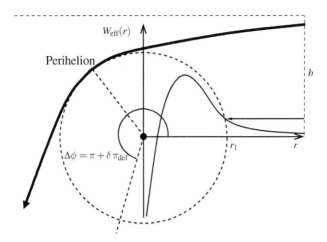

Figure 9.5 Orbit of a photon in a central potential, according to General Relativity. Superimposed is the effective potential $W_{\text{eff}}(r)$ governing the radial motion. The photon comes in from infinity with impact parameter b, reaches a minimum distance r_1, and then returns to infinity, while the polar angle changes by $\Delta\phi$. The difference between $\Delta\phi$ and π is the deflection angle $\delta\phi_{\text{def}}$.

It can happen that the impact parameter is not large compared with the Schwarzschild radius, and for black holes, this case gives rise to extraordinary effects. Some of these effects are discussed in Chapter 11.

Problems

9.1 Show that one can always reparametrise a curve in Euclidean space with coordinates $x^i(\lambda)$ such that the length of the tangent vector is constant, i.e. such that $d(\mathbf{t}^2)/d\lambda = 0$.
[Hint: what transformation $\lambda'(\lambda)$ is required so that $(d\mathbf{x}/d\lambda)^2$ is constant?]

9.2* Verify that equation (9.18) can be rearranged to give equation (9.19), the crucial step in the derivation of the geodesic deviation equation.

9.3 Use the results for the Riemann tensor outside a body of mass M obtained in Section 8.2 to verify that the geodesic deviation equation correctly predicts that two non-relativistic test particles freely falling towards the centre of a massive body move towards one another with acceleration sGM/R^3, where R is the distance of the particles from the centre of mass and s is their separation from each other, assumed much less than R.

9.4*

a) Show that the radius of a circular orbit of a massive test particle with specific angular momentum ℓ is

$$r_\circ = 3r_S \left(1 + \sqrt{1 - 3r_S^2 c^2/\ell^2}\right)^{-1},$$

where r_S is the Schwarzschild radius of the central body.

b) Find the relationship between the period T (defined as the time to rotate by 2π around the body) and radius $r_\circ$ for a circular orbit in a Schwarzschild space–time, as measured by (i) an astronaut in this orbit and (ii) a distant observer.

9.5   Calculate the approximate distance of closest approach for a photon with impact parameter b, to first order in r_S/b, assuming that $b \gg r_S$.

9.6*   In Problem 8.3, we looked at the Schwarzschild–de Sitter solution, which has a space–time interval given by

$$ds^2 = -B(r)c^2 dt^2 + B^{-1}(r)dr^2 + r^2 d\theta^2 + r^2 \sin^2\theta \, d\phi^2,$$

where $B(r) = 1 - r_S/r - \Lambda r^2/3$.

a) Calculate the effective potential $V_{\text{eff}}^{\text{GR}}(r)$ for the radial motion of a massive test particle in this metric. Sketch the potential you find for small cosmological constant Λ and $\ell \gg r_S c$, demonstrating that stable orbits exist. What happens to the orbits as Λ increases?

b) The cosmological constant is measured to be about 1.1×10^{-52} m^{-2}. Estimate the outward acceleration from the Sun caused by the cosmological constant at the Earth's radius, and compare it to the inward acceleration due to the ordinary Newtonian potential. Repeat for a star orbiting at a distance of 30 kiloparsecs outside a galaxy of mass $10^{11} M_\odot$.

Chapter 10

Tests of General Relativity

General Relativity makes three key predictions within the Solar System · these distinguish it from Newton's theory · it passes all three with flying colours · it can be tested in several other ways

According to General Relativity, the metric around gravitating bodies like the Sun and the Earth should be well approximated by the Schwarzschild solution. The masses of the planets are much less that that of the Sun, and the Sun is to a good approximation spherically symmetric. Armed with our analysis of the motion of massive and massless particles in the Schwarzschild metric, we can discuss the three most important predictions that General Relativity makes for motions within our Solar System. Two of these were proposed by Einstein, and the third is the source of the most accurate yet confirmation of predictions by General Relativity. We also review other tests of General Relativity, both in the laboratory and outside our Solar System.

10.1 Precession of Mercury's perihelion

A perfectly Newtonian orbit where a single planet orbits the Sun closes on itself. This means that the angle of closest approach to the Sun, called perihelion, is always the same. The presence of other planets in the Solar System spoils this by perturbing the gravitational field. In the case of Mercury, analysing all the effects perturbing the orbits, using Newton's theory, gives a drift in the perihelion of about 532 arcseconds per century, once the effect of the precession of the Earth's axis of rotation is removed. However even around Einstein's time the measured rate was about 574 arcseconds per century. The discrepancy of 43 arcseconds per century, with about 1 arcsecond per century uncertainty, was unexplained.

Introducing General Relativity, First Edition. Mark Hindmarsh and Andrew Liddle.
© 2022 John Wiley & Sons Ltd. Published 2022 by John Wiley & Sons Ltd.
Companion website: www.wiley.com/go/hindmarsh/introducingGR

As we stated in Chapter 9, in General Relativity even orbits of a single planet do not perfectly close, giving rise to an intrinsic perihelion shift which we will see is larger the closer the object is to the Sun. Even for weak gravity and low velocities, by which we mean small values for GM/rc^2, ℓ^2/r^2c^2 and $\mathscr{E}/c^2$, we will see that the corrections to the Newtonian expression, while small, are important.

We first set up the equation for $d\phi/dr$, in a similar way to the Newtonian case,

$$\frac{d\phi}{dr} = \frac{d\phi/d\tau}{dr/d\tau} = \frac{\ell}{r^2} \frac{1}{\sqrt{2\left(\mathscr{E} - V_{\text{eff}}^{\text{GR}}(r)\right)}}, \tag{10.1}$$

where the effective potential $V_{\text{eff}}^{\text{GR}}$ is given in equation (9.29).

We wish to integrate equation (10.1) between the turning points of the radial motion, to obtain the change in angle $\Delta\phi$. The integral can be performed by expanding both the integrand and the limits in powers of r_{S}/r (for details see Advanced Topic A2). To first order, the limits r_1, r_2 are the solutions to the equations

$$r_1 + r_2 = \frac{GM}{|\mathscr{E}|}, \quad r_1 r_2 = \frac{\ell^2}{2|\mathscr{E}|}. \tag{10.2}$$

Writing $\Delta\phi = \pi + \delta\phi_{\text{prec}}/2$, where $\delta\phi_{\text{prec}}$ is the anomalous precession angle, it is found that

$$\delta\phi_{\text{prec}} = \frac{3\pi}{2} \frac{r_{\text{S}}}{a(1-\varepsilon^2)}, \tag{10.3}$$

where $a \equiv (r_1 + r_2)/2$ is the semi-major axis of the orbit, $\varepsilon = |r_1 - r_2|/(r_1 + r_2)$ is the orbital eccentricity, and we see once again the Schwarzschild radius r_{S}.

One can rewrite this equation in terms of the more astronomically relevant orbital period T which we obtain from Kepler's Third Law (see Advanced Topic A2, equation (A2.17))

$$T^2 = \frac{4\pi^2 a^3}{c^2 r_{\text{S}}}. \tag{10.4}$$

The relativistic contribution to the perihelion shift is then

$$\delta\phi_{\text{prec}} = 6\pi \left(\frac{2\pi}{cT}\right)^2 \frac{a^2}{1-\varepsilon^2}. \tag{10.5}$$

The relevant orbital parameters for Mercury are in Table 10.1. Substituting their values into equation (10.5), we arrive at a prediction that the General Relativistic contribution to the precession of its perihelion should be 42.98 arcseconds

Symbol	Parameter	Value
a	Semi-major axis	57 909 050 km
ε	Eccentricity	0.205 63
T	Orbital period	87.9691 d

Table 10.1 The orbital parameters for Mercury relevant for the evaluation of the relativistic perihelion shift equation (10.5).

Cause	$\delta\,\phi_{prec}$ (arcseconds per century)
Venus	277.86 ± 0.68
Earth	90.04 ± 0.08
Mars	2.54
Jupiter	153.58
Saturn	7.30 ± 0.01
Uranus	0.14
Neptune	0.04
Solar oblateness	0.01 ± 0.02
Total non-relativistic	531.51 ± 0.69
Observed	574.10 ± 0.65
Difference	42.59 ± 0.94
General Relativistic precession	42.98

Table 10.2 Non-relativistic contributions to the perihelion precession of Mercury, with the precession of the equinox removed. The relative uncertainly in the General Relativity value is about 1 part in 10^5. (Data from G. M. Clemence 1947.)

per century (see Problem 10.1).

Table 10.2 gives the sources of perihelion precession, compared to the value from detailed observations of Mercury made over several hundred years. The difference between the measured value and those contributed by known non-relativistic sources is approximately (43 ± 1) arcseconds per century.

The General Relativistic prediction is in stunning agreement with the deficit in the perihelion precession rate calculated in the Newtonian model. When Einstein first calculated this in the weeks before publishing his theory in its final form, he became certain that his new General Theory of Relativity was correct.

10.2 Gravitational light bending

In Chapter 9, we studied the orbit of a massless particle such as a photon in the space–time around a massive object. We stated that light is deflected as it passes by the object. In this section, we will calculate the deflection angle.

As the photon comes in from infinity, passes by the object, and retreats to infinity, the angular coordinate changes by an amount $\Delta\phi$ which we will calculate from the geodesic equation. The deflection angle is the difference between $\Delta\phi$ and π, the angle subtended at the origin by an infinite straight line.

$$\delta\phi_{\text{def}} = \Delta\phi - \pi. \tag{10.6}$$

This angle can be computed by taking the ratio of the equations for $dr/d\lambda$ and $d\phi/d\lambda$, derived in Advanced Topic A1. We can calculate the angular coordinate change by integrating

$$\frac{d\phi}{dr} = \frac{d\phi/d\lambda}{dr/d\lambda} = \frac{1}{r^2}\left[\frac{1}{b^2} - \frac{1}{r^2}\left(1 - \frac{r_S}{r}\right)\right]^{-1/2}. \tag{10.7}$$

The limits of the integration are $r = \infty$ and the distance of closest approach $r = r_1$, which is a solution to

$$\frac{1}{b^2} - \frac{1}{r_1^2}\left(1 - \frac{r_S}{r_1}\right) = 0. \tag{10.8}$$

The result is that the General Relativistic prediction for the deflection angle of the photon is

$$\delta\phi_{\text{def}} \simeq 2\frac{r_S}{b}. \tag{10.9}$$

Actually, if one treats the photon as a ballistic particle of velocity c, a deflection can be obtained in Newtonian gravity, which is precisely half the General Relativity value. The extra contribution in General Relativity can be traced to the light responding to the spatial curvature as well as the analogue of the Newtonian potential.

The Schwarzschild radius of the Sun is 2.95 km, while the closest an observable right ray can approach is the Solar radius, 6.96×10^5 km. A light ray which just grazes the Sun ($b = R_\odot$) is therefore deflected by an angle of 1.75 arcseconds. Ordinary optical observations from the surface of the Earth cannot resolve angular differences much less than this, due to the blurring effect of the Earth's atmosphere. Hence, the effect can be seen only in light rays passing very close to the surface of the Sun. However, the Sun itself is blindingly bright. Fortunately, during Solar eclipses the Sun is shielded and it becomes possible to study stars behind it.

When Einstein published the final version of his theory, it was not known whether light was deflected or not, and so this gave a prediction which cried out to be tested. A famous joint expedition in 1919 to make observations in Brazil and the equatorial island of Príncipe sought to confirm this, and success was claimed, though there is some debate about the accuracy of the results, and whether the Newtonian value could really be ruled out on the basis of the observations. In any case, the result received wide publicity and launched Einstein's career as a media celebrity. The prediction is by now confirmed to much better than 0.03% accuracy in interferometric observations of extragalactic radio sources, strong confirmation of Einstein's prediction.

The effects of gravitational light bending can also be seen around the centres of distant galaxies and clusters of galaxies. The light being bent comes from yet more distant galaxies. If the more distant object were to be precisely aligned with the centre of mass of an intervening object, and at the right distance, its light could travel to the Earth around multiple paths, and produce a ring of images known as a **Khvolson–Einstein ring**[1] (see Problem 10.3). In practice, the alignment is never precise, and the ring is never quite complete. An image of a distant galaxy distorted and magnified by a nearer cluster of galaxies is on the cover of this book, a beautiful example of a near-complete ring. The images of distant galaxies are in fact all distorted and magnified by the gravitational field of intervening matter to some extent, and the phenomenon in all its manifestations is called **gravitational lensing**.

10.3 Radar echo delays

There was then a long interval in work to test General Relativity, partly due to a conviction that the theory was correct, and partly due to the lack of technology. In 1964, Irwin Shapiro showed that a radar signal reflected from a planet will be delayed if the signal passes near a massive body. The time delay is related to the bending of light rays; the bent trajectories are longer, and so it takes a longer time for the light to travel past the Sun (see Figure 10.1).

In order to calculate the light travel time from Earth, we will need $dr/d\lambda$ and $dt/d\lambda$ from our study of photon orbits in Advanced Topic A1. Taking the ratio, we find

$$\frac{dt}{dr} = \frac{1}{cb}\left(1 - \frac{r_S}{r}\right)^{-1}\left[\frac{1}{b^2} - \frac{1}{r^2}\left(1 - \frac{r_S}{r}\right)\right]^{-1/2}, \qquad (10.10)$$

where b is the impact parameter of the geodesic. The distance of closest approach r_1 given by the solution to

$$\frac{1}{b^2} = \frac{1}{r_1^2}\left(1 - \frac{r_S}{r_1}\right). \qquad (10.11)$$

---

[1] Orest Khvolson's name is often omitted to leave **Einstein ring**, despite his being the first to describe the phenomenon.

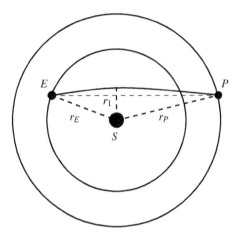

Figure 10.1 Einstein's photon has further to go. Radar signals between Earth E and another planet P are delayed as they pass the Sun S. The delay is mostly due to same effect which causes the light to bend: in the Schwarzschild metric the null geodesics along which the photons travel are curved and longer.

The light travel time from the Earth to the point closest to the Sun is the integral of equation (10.10) between r_E and r_1.

As before, we can perform the integral in an expansion in powers of r_S/r, to obtain the light travel time from the Earth to the point closest to the Sun as

$$\Delta t_E \simeq \frac{1}{c}\left(r_E + r_S \ln \frac{2r_E}{r_1} \right). \tag{10.12}$$

The first term in the brackets is just the travel time in Minkowski space–time, in the approximation that $r_E \gg r_1$. The General Relativistic time delay is the second term.

The total travel time for a signal to travel from the Earth to the planet and back is $\Delta t = 2(\Delta t_E + \Delta t_P)$, where Δt_P is the light travel time between the distance of closest approach and the planet at radius r_P. The accumulated time delay is therefore

$$\delta t_{\text{del}} = \frac{2r_S}{c} \ln \left(\frac{4r_E r_P}{r_1^2} \right). \tag{10.13}$$

This is the equation derived by Shapiro in 1964.

It turns out that reflected pulse of radar signals from a planet is too weak for an accurate determination of the light travel time. Instead it is better to use a transponder on a space probe, which can send back a much more powerful signal.

In either case there is another source of delay, which is the (frequency-dependent) slowing down of light in the medium of the Solar corona. The experiment was first successfully performed in 1976 using the Viking Mars orbiter. In 2003, the Cassini probe to Saturn provided a much more accurate check, as it was able to compensate for the Solar corona by operating at two different frequencies. Saturn is also more distant that Mars, and there was also some good fortune that the line of sight to Saturn passed very close to the Sun ($r_1 = 1.6R_\odot$) during the lifetime of the mission.

The result for the time delay was in very accurate agreement with General Relativity, at the 0.001% level. This is a very stringent test for any theory of gravity attempting to extend or replace General Relativity.

10.4 Gravitational redshift

As we first discussed in Chapter 4, the Einstein equivalence principle implies that the frequency of light should decrease as it climbs out of a gravitational potential well, by a small amount

$$\frac{\Delta f}{f} = -\frac{\Delta \Phi}{c^2}, \tag{10.14}$$

where $\Delta \Phi = gh$ is the gravitational potential difference between the places with a height difference h, and g is the acceleration due to gravity. This is a very small shift by terrestrial measures; the fractional frequency change for light travelling from the Earth's surface to infinity is $\Delta f/f \simeq -6.7 \times 10^{-10}$. Gravitational redshifts are often quoted in terms of an equivalent relative speed required to give the same Doppler shift, $\Delta f/f = v_{gr}/c$. In these terms, the gravitational redshift at the surface of the Earth is $v_{gr} \simeq 0.2\,\mathrm{m\,s^{-1}}$.

The first attempts to measure the gravitational redshift started in the 1920s, using the spectral lines of light emitted from the surface of White Dwarf stars. A White Dwarf is a star of about the mass of the Sun, which near the end of its life cycle has collapsed to a radius about that of the Earth. The gravitational redshift from the surface of a White Dwarf is therefore expected to be a factor $M_\odot/M_\oplus \simeq 3 \times 10^5$ larger than that from the Earth.

Early claims of a measurement in the spectral lines of Sirius B in 1925 are now discounted, and the first definitive measurement came in 1954, by Daniel Popper, from observations of 40 Eridani B. A gravitational redshift of $21 \pm 4\,\mathrm{km\,s^{-1}}$ was measured, against the General Relativistic prediction of $17 \pm 3\,\mathrm{km\,s^{-1}}$. The uncertainty in the prediction comes from uncertainties in the determination of the mass and radius of 40 Eridani B, which have to be inferred from precise measurements of the orbits of the star and its companions, and of the surface temperature.

For assessing the challenge faced by laboratory experiments, it is useful to look the fractional frequency change per metre of elevation at the Earth's surface, which is

$$\frac{d}{dh}\frac{\Delta f}{f} = -\frac{g}{c^2} \simeq -1.1 \times 10^{-16}\,\mathrm{m}^{-1}. \tag{10.15}$$

In order to measure the effect on Earth, one must be able to determine frequency changes very accurately.

Let us consider in detail how light is emitted and absorbed. If a quantum system like an atom or a nucleus is in an excited state, it decays with a half-life $\tau_{1/2}$, emitting a photon with frequency $f_0 = E/h$, where E is the energy difference between the initial and final states in the rest frame, and h is Planck's constant. This energy will have an uncertainty of order $\Delta E \simeq h/\tau_{1/2}$, and so the photons will be emitted with a distribution of frequencies with half-width $\Delta f = 1/\tau_{1/2}$. Likewise, a system in its ground state will absorb a photon if it is incident with frequency within about Δf of the transition frequency f_0. This gives an estimate of the height difference needed to measure a frequency shift, as the absorption probability falls to 50% due to the gravitational redshift at a height

$$h_{1/2} = \frac{4.18\,\mathrm{MeV}}{E}\frac{\mu s}{\tau_{1/2}}\,\mathrm{m}. \tag{10.16}$$

Hence, we should look for atoms or nuclei with large values of $f_0\tau_{1/2}$. Robert Pound and Glen Rebka suggested using ^{57}Fe, which has an excited nuclear state emitting a γ-ray with energy $E_\gamma = 14.4\,\mathrm{keV}$ and half-life $0.10\,\mu s$, giving $h_{1/2} = 2.9\,\mathrm{km}$.

There is a major obstacle in experiments of this kind, which is that atoms recoil when photons are emitted, by conservation of momentum, and so the emitted photon is redshifted by an amount $\Delta f/f = E/2Mc^2$, where M is the mass of the recoiling object. For an iron atom recoiling from a 14.4 keV photon, this fractional change is about 10^{-6}, far larger than that achievable by a height difference.

The key breakthrough came in 1958 when Rudolf Mössbauer discovered that when the decaying atom is in a crystal, the recoil momentum is sometimes absorbed by the lattice as a whole. This vastly multiplies the effective mass of the recoiling object, and renders the recoil effect negligible. For photons emitted via the Mössbauer effect, the main systematic to be corrected for is the Doppler shift due to the thermal vibrations of the atoms.

Exploiting the Mössbauer effect, Pound and Rebka were able to measure the gravitational redshift using the 14.4 keV line in ^{57}Fe, with source and detector (also a crystal of ^{57}Fe) separated by a height difference of 22.5 m. By exchanging source and detector, they could effectively double the height difference. They found that the ratio of the experimentally measured frequency shift Δf_{ex} to the one predicted by General Relativity Δf_{GR} was

$$\frac{\Delta f_{ex}}{\Delta f_{GR}} = -1.05 \pm 0.10, \qquad (10.17)$$

where the error was mostly statistical.

The most accurate modern experiment uses a Hydrogen maser in a satellite orbiting at a height of around 10 000 km, which has verified the General Relativity prediction to a precision of about 7×10^{-5}.

10.5 Binary pulsar PSR 1913+16

The best tests come from the strong gravity environment, rather than the weak gravity of our own Solar System. The most successful have been studies of the binary pulsar PSR 1913+16, which now has more than 40 years of accumulated measurements on its orbital properties, ultimately earning discoverers Hulse and Taylor the Nobel Prize. As well as testing General Relativity as applied to orbits, it provides convincing evidence of loss of energy by radiation of gravitational waves.

PSR 1913+16 consists of a pair of neutron stars, both of mass around $1.4 M_\odot$. One of them is a pulsar, a class of objects discovered by Jocelyn Bell Burnell, which means that it emits a very regular series of radio pulses towards us. The pulsar is in effect an extremely accurate clock in an approximately Keplerian orbit around the centre of mass of the binary system. The system is therefore an ideal laboratory for testing the predictions of General Relativity.

The observational data consists of the arrival times of the pulses. Assuming that, in the frame moving with the pulsar, the pulses are emitted at regular times T_e, one can use General Relativity to predict the arrival times at the Earth, t_a, taking into account all the Newtonian and General Relativistic corrections.

The principal correction to the arrival time is due to the difference in light travel time at different points on the pulsar's orbit, called the Rømer time delay, after the Danish astronomer Ole Rømer who used time delays within the Solar System to measure the speed of light in the 17th century.

The time delays from this largest correction depend on the projection of the pulsar orbit's semi-major axis a_1 onto the line of sight and the pulsar's orbital eccentricity ε. Some geometry based on Figure 10.2 and the equation of an ellipse (9.27) shows that the Rømer delay is

$$\Delta_R = x \frac{(1 - \varepsilon^2) \sin(\phi + \omega)}{1 + \varepsilon \cos \phi}, \qquad (10.18)$$

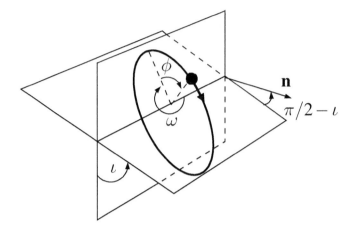

Figure 10.2   Sketch of the orbit of the pulsar, showing the angles ι (orbital inclination), ω (longitude of periastron), and ϕ (the orbital angle of the pulsar). The unit
vector **n** points to the centre of mass of the Solar System. The two planes are the
plane of the sky, orthogonal to **n**, and the plane of the orbit.

where $x = (a_1/c)\sin\iota$, ι is the orbital inclination, ϕ is the usual orbital angle (see
Section 9.4), and ω is the angle between periastron and the intersection of the
orbital plane with the plane orthogonal to the line of sight, passing through the
centre of the ellipse. This angle is called the longitude of periastron. The Rømer
time delays are periodic with the same period as that of the binary P_b. All these
parameters can therefore easily be measured. However, at this level of accuracy,
one cannot disentangle the semi-major axis from $\sin\iota$, and so one cannot measure
the total mass of the system by using Kepler's Third Law.

General Relativity predicts that the periastron should precess, in the same way
that the perihelion of Mercury precesses, meaning that there is a slow increase in
ω. The calculation of the precession is rather harder for two nearly equal mass
bodies in orbit around each other than in the case of Mercury where one mass
dominates. But if the orbits are approximately Keplerian, it is possible to show
that the periastron shift is still given by equation (10.5), with the parameter a
replaced by the semi-major axis of the relative orbit, which is connected to the
total mass of the system M through Kepler's Third Law. The periastron precession is quantified by the average rate of change of ω, which can be derived from
equation (10.5) as

$$\langle\dot{\omega}\rangle = \frac{2\pi}{P_b}\left(\frac{2\pi GM}{c^3 P_b}\right)^{2/3}\frac{3}{1-\varepsilon^2}.\qquad(10.19)$$

The total mass of the system is therefore accessible through the measurement of the periastron advance.

The next most important time delay, the so-called Einstein time delay, is caused by a combination of the gravitational redshift in the potential of the companion star, and the second-order Doppler effect from the pulsar's orbital motion. One can show that (see Problem 10.6) the time kept by the pulsar 'clock' T is related to the time coordinate in the centre of mass frame t by

$$\frac{dT}{dt} = 1 - \delta - \frac{Gm_2(m_1 + 2m_2)}{c^2 r(m_1 + m_2)}, \tag{10.20}$$

where m_1 is the mass of the pulsar, m_2 the companion mass, and r is the distance between them. The remaining term δ is mainly due to the gravitational redshift of the radio pulse originating at the pulsar surface, and is a constant.

With a knowledge of the solution of the Keplerian orbits, equation (10.20) can be integrated (see Problem 10.7) to give

$$T = t(1 + \delta') + \Delta_E, \tag{10.21}$$

where δ' is an unimportant constant and the Einstein time delay is

$$\Delta_E = -\gamma \frac{\sin \phi}{1 + \varepsilon \cos \phi}. \tag{10.22}$$

The parameter γ is given by

$$\gamma = \frac{Gm_2(m_1 + 2m_2)}{c^2 a(m_1 + m_2)} \frac{P_b \varepsilon}{2\pi}. \tag{10.23}$$

One can estimate that this effect is of order v/c smaller than the Rømer time delay, where $v \simeq \sqrt{GM/a}$ is the average orbital speed of the pulsar. The measurements give $v/c \sim 10^{-3}$, confirming that the effect is small. It is accurately measurable given the amount of data collected over the years, and enables very precise estimates of the individual masses of the two objects.

The next most important effect is the Shapiro time delay Δ_S, incurred as the radio pulses travel through the space–time near the pulsar's companion. By revisiting the discussion of the Shapiro delay in the Solar System, and allowing for the fact that the distance to the Earth d_e is much larger than the distance of the pulsar from the centre of mass of the binary system, one can show that

$$\Delta_S = \frac{2Gm_2}{c^3} \ln \frac{2d_e}{r + \mathbf{n} \cdot \mathbf{r}}, \tag{10.24}$$

where **r** is the displacement between the pulsar and its companion, and **n** is the unit vector in the direction of the Earth. A geometrical exercise similar to that involved in the calculation of the Rømer time delay shows that the part of the Shapiro time delay which varies in time, and therefore can more easily be measured, is

$$\Delta_S = -\frac{2Gm_2}{c^3} \ln\left(\frac{1 + \varepsilon\cos\phi}{1 - \sin\iota\sin(\phi+\omega)}\right). \tag{10.25}$$

The Shapiro time delay is of order v^2/c^2 smaller than the Rømer time delay, and has been measured only recently, due to the pulsar becoming more favourably aligned with its companion as its periastron precesses. The measurement of the Shapiro time delay gives no new information about the parameters of the system, as they are already measured, but it does provide a valuable test of General Relativity in a system with strong gravitational fields.

The final General Relativistic effect is the aberration time delay Δ_A. The source of the pulse is thought to be a beam coming from a patch on the surface of the rapidly rotating neutron star, which we observe if it is aligned with the direction **n**, a vector defined in the binary's centre of mass frame. In the frame of the neutron star, which is moving with velocity **v**, the beam is observed only if it is aligned at a slightly different direction, **n'**. The angle between the two directions is $\theta = |\mathbf{n} \times \mathbf{v}|/c$. The resulting delay is of order $(v/c)P_p$, where P_p is the pulsar period. This turns out to be of the same order of magnitude as the Shapiro time delay, and has not yet been measured.

Perhaps the most exciting prediction of General Relativity is that the binary should emit gravitational waves, as we explore in detail in Advanced Topics A3 and A4. The resulting loss of energy means that the neutron stars get closer, and the period of their mutual orbit decreases. The measured rate of period decrease $\dot{P}_b$ is in excellent accord with the General Relativistic prediction.

In Table 10.3, we list the most recent determination of the orbital parameters we have discussed, showing the astonishing accuracy with which they can be determined. From these parameters it can be deduced that the masses of the pulsar and its companion are $(1.438 \pm 0.001)M_\odot$ and $(1.390 \pm 0.001)M_\odot$.

The knowledge of the masses of the neutron stars allows a prediction for the rate of decrease of the period due to the energy loss through gravitational radiation, using the formula given in Advanced Topic A4. The General Relativity prediction is

$$\dot{P}_b^{gr} = -(2.402\,63 \pm 0.000\,05) \times 10^{-12}, \tag{10.26}$$

which should be compared with the 'intrinsic' period derivative $\dot{P}_b^{intr}$, which is the value of the period derivative after correction of the observed value $\dot{P}_b^{obs}$ by the

Parameter	Value
$x \equiv (a_1/c)\sin\iota$ (s)	2.341 776(2)
ε	0.617 134 0(4)
P_b (d)	0.322 997 448 918(3)
$\langle \dot{\omega} \rangle$ (deg/yr)	4.226 585(4)
γ (μs)	4.307(4)
$\dot{P}_b^{\text{obs}}$	$-2.423(1) \times 10^{-12}$
$\dot{P}_b^{\text{intr}}$	$-2.398(4) \times 10^{-12}$

Table 10.3 Orbital parameters of PSR 1913+16, as determined from pulse arrival times. The bracketed number shows the uncertainty on the final digit. (Data from J. M. Weisberg and Y.¶. Huang 2016.)

relative accelerations of the Solar System and the binary pulsar in the Galactic frame. The agreement is extraordinary, and gives strong support to the existence of gravitational waves in General Relativity even without their direct observation.

10.6 Direct detection of gravitational waves

The gravitational waves just discussed are ripples in space–time that are generated by rapidly orbiting compact objects like neutron stars, and then propagate freely through the vacuum. Their existence was predicted by Einstein in 1916 as solutions to the vacuum Einstein equations. The observations of the binary pulsar discussed in Section 10.5 have provided longstanding evidence for their reality as the explanation for that system's long-term energy loss. The Universe should be suffused with these waves, like the patterns on a pond on a rainy day, the Earth being constantly — but very, very gently — buffeted by their passage.

One of the most remarkable recent results across all science was the direct detection of gravitational waves from distant astronomical objects, specifically systems of merging black holes and neutron stars. It was announced in 2016 that in September of the previous year the Laser Interferometer Gravitational-wave Observatory (LIGO), two huge detectors in widely separated locations in the United States, had unambiguously detected ripples in space–time emitted during the merger of two black holes over a billion years ago. This first event was named GW150914, the signal matching the form expected if the two black holes each had a mass around 30 Solar masses. This landmark was only possible through decades of development of sensitive laser interferometers capable of measuring displacements smaller than one part in 10^{21} across distances of kilometres.

The merging of close compact binaries takes place because the gravitational waves carry away energy, causing the orbit to become ever more tightly bound until collision of the objects takes place. We give a detailed account in Advanced

Topic A4. As the objects spiral ever closer, their orbital frequency goes up and hence so does the frequency of the emitted gravitational waves. This leads to a distinctive 'chirp' pattern, rising sharply only within the last second before the actual merger — you can see an example in Figure A4.3. Accurate prediction of the signal is a challenging task requiring advanced numerical methods, but its high information content brings both detailed information of the properties of the merging system and tests of the General Relativity calculations used to predict the signal.

The discovery heralds a new epoch of **multi-messenger astronomy**, where new knowledge about astrophysical objects can be obtained by simultaneous observations in both electromagnetic radiation and gravitational waves. The first result of this kind is already striking; the coincident arrival of gravitational waves and photons from source GW170817, a merger of two neutron stars, shows that gravitational waves propagate at the speed of light. This is predicted by General Relativity and demonstrated to an accuracy better than one part in 10^{15}, and strongly constrains or eliminates some proposed alternatives to General Relativity.

Gravitational waves have thus been thrust to the forefront of gravitational research, with ambitious proposals for ground- and space-based observatories and bringing new energy to theoretical investigations and modelling of complex high-gravity astronomical environments. We have devoted Advanced Topics A3 and A4 to a detailed introduction to gravitational wave theory and detection.

Problems

10.1   Verify, from equation (10.5) and the orbital parameters for Mercury given in Table 10.1, that the General Relativity prediction for its perihelion precession is indeed 43 arcseconds per century. If the Sun were to collapse to form a white dwarf with radius $R_{\mathrm{wd}} = 10\,000$ km, without losing mass or disturbing the orbit of Mercury (rather unrealistic assumptions), what would the perihelion precession be?

10.2   In Problem 4.1, you calculated the observed frequency of light emitted with frequency f_{em} at a radius $r = GM/c^2$ from a body of mass M. Explain why this calculation is wrong. Show that the correct formula for the gravitational redshift at infinity from a stationary source at distance r from a spherically symmetric, non-rotating body of mass M is

$$f_\infty = f_{\mathrm{em}} \left(1 - \frac{2GM}{rc^2} \right)^{1/2} .$$

10.3* Consider a point light source S, a non-rotating central body L, and an observer O, all in a line. The distance between the observer and source is D_s, the observer and lens is D_l, and between the lens and the source D_{ls}. Assume that the distances are much larger than the impact parameter of the light rays, which in turn is much larger than the Schwarzschild radius of the central body. Show that the observer sees the source as a ring of angular diameter

$$\theta = \sqrt{\frac{2r_s D_{ls}}{D_l D_s}}\,.$$

[Hint: you can assume that all the deflection takes place in a plane containing the central body, and orthogonal to the line OLS.]

10.4

a) Show that the path a photon takes in the time-delay experiment is indeed longer than the Euclidean distance between Earth and the planet.

b) In Newtonian dynamics, where the photon is treated as an ordinary particle moving at speed c, the deflection angle is one-half the value predicted by General Relativity. Assuming that the deflection is concentrated near the Sun, what would the time delay be for this Newtonian particle, to first order in r_S/b?

10.5* Verify that the differences in the light travel time of pulses from a pulsar in an elliptical orbit, the Rømer time delay Δ_R, is indeed given by equation (10.18). [Hint 1: refer to Section A2.1 on Keplerian orbits, where Figure A2.1 shows how to set up Cartesian coordinates with origin at the centre of the ellipse, and defines useful angles. You will need equation (A2.19).]
[Hint 2: first consider the case where $\omega = 0$, and then make a rotation in the orbital plane.]

10.6* Derive the relationship between the time in the pulsar frame T and the pulsar system's centre of mass frame t given in equation (10.20). This calculation is best carried out in the weak-field approximation to the Schwarzschild metric in the isotropic coordinates of Problem 8.2, repeated here for convenience:

$$ds^2 = -c^2 dt^2 (1 + 2\Phi) + (1 - 2\Phi)(d\tilde{x}^2 + d\tilde{y}^2 + d\tilde{z}^2).$$

[Hint 1: the pulsar is a moving clock in this frame, with speed $v = |d\tilde{x}/dt|$.]
[Hint 2: you can assume a Keplerian orbit as a first approximation.]

10.7   Integrate equation (10.20) in order to derive the Einstein time-delay formula (10.21).

[Hint: it will again pay off to study Section A2.1 on Keplerian orbits, where efficient changes of variable are given. Look out for the eccentric anomaly and equation (A2.19).]

Chapter 11

Black Holes

Black holes!!

In this chapter, we return to the properties of the Schwarzschild space–time, now including the strong-gravity regime at small radius. We introduce and discuss singularities of the metric, assisted by studying light travel in the Schwarzschild space–time. The singularities prove to be key in physically interpreting the unusual properties of the space–time. We end the chapter with a description of orbits close to the Schwarzschild radius, and the many roles that black holes may play in astrophysics and cosmology.

11.1 The Schwarzschild radius

The Schwarzschild metric is

$$ds^2 = -c^2 \left(1 - \frac{2GM}{rc^2} \right) dt^2 + \left(1 - \frac{2GM}{rc^2} \right)^{-1} dr^2 + r^2 d\theta^2 + r^2 \sin^2 \theta \, d\phi^2 .$$

(11.1)

It gives the space–time outside a spherically symmetric body of mass M.

An important feature of the metric is that it is singular, meaning that the metric coefficients go to infinity, at both $r = 0$ and $r = 2GM/c^2$. Such **singularities** will play a vital role in what follows. We recall that the latter is known as the **Schwarzschild radius**, which we denote r_S. Some approximate values of the Schwarzschild radius, for objects of various characteristic masses, are given in Table 11.1.

In all cases, the object in question is bigger than the Schwarzschild radius. As the Schwarzschild solution is the vacuum solution that holds only outside whatever object is sourcing the gravitational field, we see that in each of these cases

Introducing General Relativity, First Edition. Mark Hindmarsh and Andrew Liddle.
© 2022 John Wiley & Sons Ltd. Published 2022 by John Wiley & Sons Ltd.
Companion website: www.wiley.com/go/hindmarsh/introducingGR

Object	Mass (kg)	Radius	r_S
Earth	6×10^{25}	6×10^3 km	9 cm
Sun	2×10^{30}	7×10^5 km	3 km
PSR J0348+0432	4×10^{30}	$10 - 15$ km	6 km
Neutron	2×10^{-27}	9×10^{-16} m	2×10^{-54} m

Table 11.1   Approximate masses, radii, and Schwarzschild radii r_S for selected astronomical bodies, with the neutron for comparison. PSR J0348+0432 is a neutron star, whose radius is not accurately known.

the solution is only relevant at radii larger than the Schwarzschild radius. Hence, there is no need to be concerned about possible singular behaviours in those cases.

The object which comes closest is PSR J0348+0432, which is a neutron star, one of the most massive for which the mass is known accurately, although neutron star radii are much less well determined. As its name suggests, a neutron star consists of closely packed neutrons, perhaps with a core of quark matter.

But what if the source material can be concentrated within the Schwarzschild radius? Indeed, this would seem to be the case for neutron stars with masses much larger than twice the mass of the Sun, since assuming that nuclear density is approximately maintained, their radius should scale in proportion to the one-third power of the mass, while the Schwarzschild radius is directly proportional to it. Such objects are known as **black holes**, because of their ability to trap light, and are the main topic of this chapter. Those properties hinge on the nature of the singularities in the Schwarzschild metric.

11.2   Singularities

A location where the metric coefficients blow up to infinity is an example of what is known as a **singularity**. We already saw that for the Schwarzschild space–time, this happens at $r = 0$ and $r = r_S$. It is very common for metrics to exhibit singularities — we will encounter another example when we consider cosmology later in the book — and interpreting them correctly is crucial. Mathematical singularities (or divergences) crop up in many branches of physics, and are normally taken as indicating that the theory being used is in some way incomplete.

There are two types of singularity which will be relevant to our discussion of the Schwarzschild space–time. One is a genuine singularity of the space–time itself, where the space–time becomes infinitely curved and our equations can no longer apply. This is known as a **curvature singularity**. The other is due to a bad choice of coordinates to try and describe a space–time which is in fact smooth and well-behaved. In the second case it is the particular coordinate system, rather than the space–time itself, which is becoming pathological. This latter type of singularity can be removed by making a wiser choice of coordinate system, and is known as a **coordinate singularity**.[1]

One hint at the nature of singularities comes from the components of the Riemann tensor. Interestingly, it turns out that some are finite at the Schwarzschild radius, e.g. R^{tr}_{tr} (see Problem 8.1), with the general form r_S/r^3. However, the precise form of the components of the Riemann tensor depends on the choice of coordinates made, and so does not give a definitive result. If instead we form invariant combinations of the tensor, we can assess the nature of the singularity independent of the coordinate choice.

For example, one can multiply the Riemann tensor by itself and contract on all indices to get a scalar quantity, sometimes referred to as the Kretschmann scalar

$$K \equiv R^{\mu\nu\rho\sigma} R_{\mu\nu\rho\sigma} = 12\frac{r_S^2}{r^6}. \tag{11.2}$$

The final expression is specific to the Schwarzschild space–time. Computing it is an algebraic challenge which you are asked to attempt in Problem 11.1. This result indicates we have a genuine and unremovable singularity in the space–time at $r = 0$.

Note that we have to make combinations of the Riemann tensor to test such infinities; the vacuum equations $R^{\mu\nu} = 0$ guarantee that any combination of the Ricci tensor and Ricci scalar always vanishes, and the Ricci tensor also encodes all the information in any possible contractions of the Riemann tensor alone (see Problem 5.13).

In contrast to its divergence at $r = 0$, the Kretschmann scalar is perfectly finite $r = r_S$. This strongly suggests that the coordinates used to display the metric in equation (11.1) possess only a coordinate singularity at the Schwarzschild radius. However we cannot yet be sure that there are not other, even more complex, invariants that do diverge at r_S. To verify the hypothesis of a coordinate singularity, we need to find alternative coordinate systems that eliminate the singularity.

11.3 Radial rays in the Schwarzschild space–time

A key to understanding the Schwarzschild space–time is the behaviour of light rays moving radially, so that $d\theta = d\phi = 0$. The null geodesic condition $ds^2 = 0$ becomes

$$c\,dt = \pm\frac{dr}{1 - r_S/r}, \tag{11.3}$$

---

[1] A more general definition of singularities uses the idea of geodesic incompleteness, where free-falling trajectories end within a finite proper time and can be predicted no further.

where the plus sign is for outgoing rays (radius increasing with time) and the minus sign for incoming rays. When we integrate this, the solutions depend both on whether the light ray is directed inwards or outwards, and on whether we are inside or outside the Schwarzschild radius. First outside, we have

$$ct \;=\; r + r_S \ln\left(\frac{r}{r_S} - 1\right) + ct_W \qquad \text{Outgoing;} \qquad (11.4)$$

$$-ct \;=\; r + r_S \ln\left(\frac{r}{r_S} - 1\right) - ct_W \qquad \text{Ingoing,} \qquad (11.5)$$

where t_W is an integration constant, the time at which the geodesic crosses a fixed radius $r \simeq 1.278 r_S$ (the value that makes the r-dependent terms in the equations equal zero). For rays inside the Schwarzschild radius, we have

$$ct \;=\; r - r_S \ln\left(1 - \frac{r}{r_S}\right) + ct_0 \qquad \text{Outgoing;} \qquad (11.6)$$

$$-ct \;=\; r - r_S \ln\left(1 - \frac{r}{r_S}\right) - ct_0 \qquad \text{Ingoing.} \qquad (11.7)$$

In this case, the integration constants t_0 are the times at which the geodesics reach or leave the singularity at $r = 0$.

These solutions are shown schematically in Figure 11.1. As you see, the trajectories do not cross the Schwarzschild radius when viewed in this coordinate system; even an ingoing light ray takes an infinite amount of time to reach the Schwarzschild radius. If you are already familiar with some of the properties of black holes and the unfortunate consequences of falling into one, this will surely come as a surprise; how can anything fall into a black hole if even an inwardly directed light ray needs an infinite amount of time to reach the Schwarzschild radius?

The resolution is that the singular coordinate system is misleading. In particular, it turns out that an *infinite* amount of time, as measured using the t coordinate, may correspond to a *finite* change in the affine parameter describing the geodesic.

In fact any freely falling particle reaches the Schwarzschild radius in a finite proper time, and since proper time is the time actually experienced by a free falling observer, it is the important property. But what happens to this observer after the proper time at which they reach $r = r_S$, which appears not to be represented at all by the t coordinate? To answer this, we need better coordinates.

Finally, another pathology of the coordinates is that inside the Schwarzschild radius r takes on the role of the time coordinate, since $g_{rr} < 0$, and t becomes a space coordinate, $g_{tt} > 0$. The time and space coordinates swap identities as we cross the Schwarzschild radius, making Figure 11.1 difficult to interpret. To help

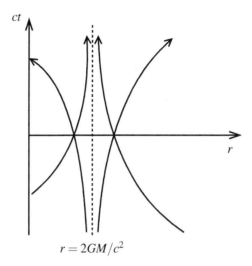

$$r = 2GM/c^2$$

Figure 11.1 A schematic illustration of the behaviour of the four solutions for radial light rays in the Schwarzschild metric. Ingoing rays inside the Schwarzschild radius $r_S = 2GM/c^2$ reach the central singularity in finite time.

understand particle trajectories in the Schwarzschild space–time, we would like to have a coordinate system so that one of our symbols is a time variable everywhere in the space–time, and the others are always spatial coordinates.

11.4 Schwarzschild coordinate systems

Unfortunately, good coordinate systems are hard to come by; it took nearly 50 years from the discovery of the Schwarzschild solution to find a fully satisfying version. Some steps along the way are the Isotropic Coordinates (see Problem 8.2) and the Eddington–Finkelstein coordinates (Problem 11.3), each of which successfully removes the singularity at the Schwarzschild radius. But best of all are the Kruskal–Szekeres coordinates.

We won't explore the transformation in detail, but for the record it is

$$u = \left(\frac{r}{r_S} - 1\right)^{1/2} e^{r/2r_S} \cosh \frac{tc}{2r_S} \,; \tag{11.8}$$

$$v = \left(\frac{r}{r_S} - 1\right)^{1/2} e^{r/2r_S} \sinh \frac{tc}{2r_S} \,, \tag{11.9}$$

for $r > r_S$ and the same with the square root sign reversed for $r < r_S$. It converts

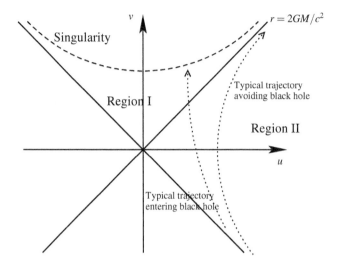

Figure 11.2 Particle motion in the Schwarzschild space–time, displayed in Kruskal–Szekeres coordinates (u, v). A particle can cross into Region I (inside the Schwarzschild radius $r_S = 2GM/c^2$) with finite coordinate values, and once there cannot avoid hitting the singularity at $r = 0$.

the metric into the form

$$ds^2 = -\frac{4r_S^3}{r}e^{-r/r_S}\left(dv^2 - du^2\right) + r^2 d\theta^2 + r^2 \sin^2\theta\, d\phi^2, \qquad (11.10)$$

where r is an implicit function of u and v given by the coordinate transformation as

$$\left(\frac{r}{r_S} - 1\right) e^{r/r_S} = u^2 - v^2, \qquad (11.11)$$

this form holding both inside and outside the Schwarzschild radius.

This coordinate system has a series of desirable properties:

1. v is a time coordinate everywhere ($g_{vv} < 0$) and u a space coordinate everywhere ($g_{uu} > 0$).

2. Radial light rays obey $dv = \pm du$. So in a space–time diagram using these coordinates, light rays travel at 45°, just as in Special Relativity.

3. Lines of constant r are hyperbolae, and constant t are straight lines through the origin.

The diagram shows the space–time in the u–v coordinate system. The most important feature is that the singularity at $r = 0$ now appears as the hyperbola $v^2 = u^2 + 1$. This corresponds to *two* space-like surfaces, one at positive values of v, and the other at negative values. We distinguish them by calling them the **future singularity** and the **past singularity**. The diagram shows only the future singularity. The idea that the centre of the black hole is at some location in space that remains fixed for all time is simply incorrect, as you might already have deduced from our discovery that the r coordinate in the Schwarzschild system is actually a time coordinate due to $g_{rr} < 0$.

In these coordinates, we see that the Schwarzschild radius $r = r_S$ is a surface with $ds^2 = 0$, which has the strange implication that outgoing radial light rays starting at the Schwarzschild radius would stay there. Typical radial trajectories will eventually cross the Schwarzschild radius, though this can be avoided by applying a constant eternal acceleration (i.e. hovering above the Schwarzschild radius forever). The dotted lines show two typical trajectories, one avoiding the black hole and one entering it. These coordinates make it clear that it is possible to cross through the Schwarzschild radius, but not back again. Having passed through the Schwarzschild radius, the singularity lies ahead of an intrepid astronaut in time, not at a particular location in space; hence the name future singularity. Moreover, the proper time between crossing the Schwarzschild radius and reaching the future singularity is finite, whatever action they may choose to take. An encounter with the singularity has become as unavoidable for our astronaut as an encounter with a forthcoming day of the week is for us. Whatever we do, we can't avoid next Tuesday.

11.5 The black hole space–time

The analysis of the Schwarzschild space–time that we have carried out, particularly in Kruskal–Szekeres coordinates, has shown the following set of properties.

1. Nothing can escape from inside a black hole. The Schwarzschild radius marks out a sphere, known as the **event horizon**, behind which events are forever hidden to any observer who remains outside the black hole. We see this directly from Figure 11.2, which shows that the event horizon cannot be crossed by any light ray travelling in the interior, denoted Region I.

2. The event horizon is not a real singularity, and observers in Region II can readily pass inside it. In doing so they experience a finite time passing. Again this is most apparent in the Kruskal–Szekeres coordinate system, though simpler transformations such as the isotropic coordinates of Problem 8.2 are already sufficient to show the fictitious nature of the Schwarzschild singularity (see Problems 11.2 and 11.3).

3. Any observer passing through the horizon will then inevitably encounter the true singularity within a finite proper time (see Problem 11.2).

4. In practice, the observer will not survive the trip all the way to the singularity. The gravitational force on whatever part of their body is closest to the singularity is greater than that on the furthest part, and the difference is experienced as a stretching force. This is the same phenomenon by which the Moon gives rise to two tides on Earth (one on the near side where its attraction is greater than the average, one on the far side where it is less), and is known as a tidal force. In General Relativity, the tidal acceleration at a point is given by the components of the Riemann tensor (see Problem 9.3). This grows without limit on the approach to the singularity, stretching and ultimately breaking any object. This inevitable destruction is sometimes referred to as 'spaghettification'!

5. A distant observer watching an astronaut heading towards a black hole sees them asymptotically approaching the event horizon, seeing their time (for instance, the beating of their heart) passing ever more slowly. As they approach their light is redshifted more and more, ultimately becoming infinitely so at the event horizon. This is because any observer outside the horizon can at best look back along a null line, which cannot have emerged from the other side of the null line that forms the horizon. Even in the external observer's infinite future, the astronaut is still seen approaching the event horizon more and more slowly.[2]

6. By contrast, the astronaut within the black hole continues to be able to observe what is happening outside, as there are plentiful light ray trajectories which cross the event horizon and intersect their course.

7. The radius of black holes is proportional to their mass. Contrast this with a solid body of fixed density, where we would need eight times the mass in order to double the radius. So black holes become less and less dense the larger they are, and the gravitational forces at the event horizon weaker and weaker. Indeed, for a big enough black hole, a spaceship might find itself to have inadvertently crossed inside the event horizon without noticing, though their fate will become clear as soon as they try to reverse their steps.

8. In Kruskal–Szekeres coordinates, one finds regions in the Schwarzschild space–time, other than Regions I and II, which do not appear in the original (t, r) coordinates. This is an example of the possibility, mentioned at

---

[2]In practice photons are discrete individual objects, so eventually the last ever photon will be emitted by the astronaut after which they have vanished from sight.

the start of Chapter 5, that a set of coordinates does not cover the complete spacetime. These regions, which include the past singularity at $v = -\sqrt{u^2 + 1}$, are not thought to be part of the space–time of a real black hole.

11.6 Special orbits around black holes

When we discussed orbits in Chapter 9, we ignored those passing close to the Schwarzschild radius as they would be inside an ordinary astronomical object. But for the black hole space–time we need to consider them, which will unveil new possibilities that have no analogue in the theory of Newtonian orbits.

Chapter 9 showed us that we can think of the radial motion of a particle in orbit as motion in an effective potential, with a conserved quantity $\mathscr{E}$ with dimensions of velocity squared, and a conserved specific angular momentum ℓ. The shape of the effective potential for large ℓ is given in Figure 9.4. Problem 9.4 shows that the potential takes this shape providing $\ell > \sqrt{3} r_S c$, with two turning points where it takes values $V_{\max}$ and $V_{\min}$.

Returning to the characterisation of orbits in terms of the constant $\mathscr{E}$, we can add the following cases:

- $\mathscr{E} = V_{\max}$: There is an unstable circular orbit. Incoming particles with precisely this energy asymptote to the circular orbit as $t \to \infty$.

- $\mathscr{E} > V_{\max}$: Ingoing solutions cross the Schwarzschild radius and hit $r = 0$. These are trajectories which would pass close to r_S if they were not deviated by the Schwarzschild space–time.

It seems that there are also possible outgoing trajectories, starting at $r = 0$. Trajectories emerging from the singularity and escaping to infinity are in fact associated with the unphysical part of the Schwarzschild space–time, sometimes called a white hole, which we mentioned at the end of Section 11.5.

When $\ell < \sqrt{3} r_S c$ the two extrema of the potential disappear (see Problem 9.4) and the ingoing particles always reach zero radius. We call such trajectories **plunge orbits**. There is only one plunge orbit in Newtonian gravity, where $\ell = 0$ and the particle is on a precisely radial trajectory. In GR all trajectories with $\ell < \sqrt{3} r_S c$ unavoidably reach $r = 0$.

There are also special orbits for photons in the Schwarzschild space–time. Again, their radial motion can be understood in terms of motion of an ordinary particle in an effective potential W_{eff}, shown in Figure 9.5, with a conserved quantity U with dimensions of velocity, and a conserved quantity ℓ with units of specific angular momentum (length times velocity).

Photon orbits are classified in terms of their impact parameter $b = \ell/U$. If this becomes sufficiently small, there is no solution to equation (10.8). The radial coordinate does not have a minimum, and carries on decreasing until it reaches zero in a finite affine parameter interval. This is a plunge orbit for the photon: it has insufficient angular momentum to prevent it from hitting the central body.

The special case separating these options is $b^2 = 1/W_{\text{max}}$. Here the photon can travel in a circular orbit at a radius where the effective potential reaches its maximum, $r_\circ = 3r_S/2$. Only a black hole is small enough for such orbits to be possible. As the orbit is at the maximum of the effective potential, it is unstable to perturbations inwards or outwards. A photon incoming from infinity at precisely this impact parameter will asymptotically approach the circular orbit, while photons at marginally greater impact parameter may suffer very large deflections or even spiral many times around the black hole before escaping back out to infinity.

The ability of a black hole to bend light around it gives rise to spectacular optical effects. For example, if one illuminates a black hole, one will see a series of concentric rings, due to light whose deflection angle is an odd multiple of π, a phenomenon known as black hole glory. Astrophysical black holes are likely to be surrounded by disks of hot gas, and from a suitable angle it is possible to see both sides of the disk on the far side of the black hole.[3] In Section 11.7 we briefly discuss the first imaging of an astrophysical black hole using the Event Horizon Telescope.

11.7   Black holes in physics and in astrophysics

Black holes offer enduring fascination both for their theoretical beauty and for their surprising relevance to the real Universe. We end by briefly describing some of these topics; if you want to learn more they are covered in detail in numerous advanced textbooks.

The Schwarzschild solution is the simplest solution describing the space–time outside an isolated object. Many more general solutions are known. One is the Reissner–Nordström space–time which describes a charged black hole (see Problem 11.4). This is not thought to be of any astrophysical relevance since any charged object will preferentially accrete oppositely charged objects to cancel out the charge, but it is of theoretical interest as a relatively simple generalisation of the Schwarzschild solution.

Rotating black holes are described by the Kerr metric, first derived by Roy Kerr in 1963 (see Problem 8.4). It is an example of a metric which cannot be written in a diagonal form. Since astrophysical objects are likely always to possess some rotation, due to angular momentum conservation as they form, this is the most appropriate solution to use where possible, though for all astronomical objects except neutron stars and black holes the Schwarzschild solution provides

---

[3] See the book *Black Holes*, edited by J.-P. Luminet, and the film *Interstellar*.

an excellent approximation. The solution shows that there is a maximum possible rotation speed for a given mass.

There is also an exact solution describing a rotating charged black hole, the Kerr–Newman metric. This is considered to be the most general black hole solution, at least in three space dimensions. Another way of saying this is that, when electromagnetism is included, all black hole solutions can be completely characterised by their mass, angular momentum, and charge. This formulation is known as the **no-hair theorem**; here, 'hair' is a picturesque description of any other physical property such as baryon number or lepton number. Once the three basic properties of a black hole are specified, they have no further distinguishing features.

In a black hole space–time there is always a curvature singularity, which is reached in finite proper time by an infalling particle (see Problem 11.5), or finite affine parameter in the case of an infalling photon. This is a deeply disturbing property to physicists, as the particle seems to cease to exist, without any explanation for how this happens. It was originally hoped that the existence of curvature singularities was an unphysical consequence of the special symmetries of black hole solutions. However this hope was dashed by Roger Penrose with his celebrated **singularity theorem**, which showed that under some very general assumptions, such as the positivity of energy, singularities are always present in a space–time with closed surfaces from which light cannot escape, like the Schwarzschild event horizon. Crucially, he did not have to make any assumptions about the symmetry of the surface, showing that singularities are a very general feature of General Relativity. His achievement was recognised by a share of the Nobel Prize in 2020.

The presence of a singularity does not spoil our ability to predict what happens outside the black hole as, except in very special cases,[4] the event horizon shields it from the view of any observer who stays outside the black hole. This observation prompted Penrose to formulate the **cosmic censorship conjecture**, which states that all singularities are hidden behind event horizons, regardless of the gravitational sources. This remains unproven, and indeed there remains active debate even on how to formulate the hypothesis as a statement that could be rigorously proven.

While popular imagination often regards black holes as exotic and speculative objects, there is plentiful evidence for the actual existence of black holes in a wide variety of astrophysical settings. The end point of evolution of a sufficiently-massive star is believed to be a black hole, due to the existence of maximum permitted masses for the alternative possibilities of white dwarfs or neutron stars. There is some debate as to just how massive the star must initially be, due to mass loss both during its normal stellar evolution and in the supernova explosion that

---

[4]See Problem 11.4 for one of them.

ends its life, but 20 Solar masses is a typical estimate resulting in a final black hole of mass 4 Solar masses or above. Black holes of approximately Solar mass might also form through a neutron star accreting material that takes it over the maximum stable mass for such objects, or via mergers of neutron stars and/or white dwarfs.

The gravitational waves featuring in the first direct detection of such waves, described at the end of Chapter 10, are believed to have been emitted during a merger of two black holes. The initial masses of the two black holes were approximately 36 and 29 Solar masses, with the post-merger remnant being about 62 Solar masses. The mass difference was radiated predominantly as gravitational waves and all within a fraction of a second, a prodigious emission power that momentarily exceeded the luminosity of the entire visible Universe.

On a much more massive scale, most or perhaps even all galaxies are believed to host super-massive black holes at their centres. In the case of our Milky Way, an extremely compact radio source named Sagittarius A* (after the constellation in which it resides) was discovered at its centre in 1974 by Bruce Balick and Robert Brown. Subsequent study of the eccentric orbits of stars around this object (made in infrared wavebands, as optical light does not reach us because of intervening dust clouds) reveal that approximately four million Solar masses are packed into a region smaller than the Earth's orbit, which can only be a black hole. Andrea Ghez and Reinhard Genzel received a share of the 2020 Nobel Prize for this discovery.

While black holes themselves are, of course, invisible, the strong gravity in their environment creates extremely energetic environments of fast-moving particles, leading to very strong emission of radiation in their immediate vicinity. In the case of the Milky Way, the black hole region is seen primarily as a strong radio source, but black holes in other galaxies can be much more powerful. Black holes are believed to power some of the most energetic sources in the Universe, providing the energy source for quasars and active galactic nuclei.

In 2019, an astrophysical black hole was directly imaged for the first time. This was achieved by the Event Horizon Telescope, a network of radio observatories across the globe whose signals are combined interferometrically to give a telescope whose effective size is that of the entire Earth. They imaged the super-massive black hole at the centre of the nearby galaxy M87, whose mass is inferred to be about 7 billion Solar masses. Of course the black hole itself is not imaged; what is seen is a dark disk at the centre of the bright emission from the material surrounding the black hole. Due to the bending of light paths by the black hole, the dark shadow's size is about 2.5 times larger than the actual black hole. Similar techniques may well have imaged our own galaxy's central black hole by the time you read this.

In 1975, Stephen Hawking showed that although material can never escape from a classical black hole, according to quantum mechanics black holes do radiate energy and evaporate. Their radiation is black-body radiation with a characteristic temperature $T \propto 1/M$. The inverse proportionality means that the small black holes are the hot ones and evaporate the quickest. It is nice to write this with all the constants of proportionality in place

$$k_B T = \frac{\hbar c^3}{8\pi GM},$$ (11.12)

as it features four of the most fundamental constants of nature: the Boltzmann constant, Planck's constant, the speed of light, and Newton's gravitational constant. This makes it one of the most beautiful equations you are ever likely to see. Black holes are also possess an entropy, which is simply proportional to the event horizon area, $4\pi r_S^2$. The full formula states that the entropy is Boltzmann's constant multiplied by one quarter of the event horizon area when expressed in Planck units (see Chapter 14). Restoring those units gives

$$S = k_B \frac{c^3}{G\hbar} \pi r_S^2 = \frac{4\pi G k_B}{c\hbar} M^2.$$ (11.13)

For black holes of stellar mass and above, Hawking evaporation is utterly negligible even over the lifetime of the Universe. For Hawking evaporation to be important, a black hole should have a mass of no more than 10^{15} g, a mass more characteristic of a mountain than of an astronomical object. There is no known way for such light black holes to be created in the present Universe, but many astronomers have speculated that there were large enough variations of the density in the extremely young Universe to allow this possibility. Such black holes are known as **primordial black holes** or PBHs. As yet there is no evidence for the actual existence of such objects, though the lack of any detection thus far does rule out some candidate scenarios for how the young Universe might have behaved.

Problems

11.1 Demonstrate by direct calculation that the scalar product of the Riemann tensor with itself in the black hole space–time is given by equation (11.2).

11.2 Recall the Schwarzschild metric in isotropic form derived in Problem 8.2. Find the values of the isotropic radial coordinate $\tilde{r}$ for which the metric is singular, and comment.

11.3* Eddington–Finkelstein coordinates change the Schwarzschild time coordinate from t to $\tilde{t}$ according to

$$c\tilde{t} = ct \pm \left(r + r_S \ln \left| \frac{r}{r_S} - 1 \right| \right),$$

with other coordinates unchanged. Find the Schwarzschild metric in the new coordinates. What feature of the new metric is desirable?

11.4 [Hard!] A black hole of mass M and electric charge Q has metric

$$ds^2 = -c^2 \left(1 - \frac{r_S}{r} + \frac{r_Q^2}{r^2} \right) dt^2 + \left(1 - \frac{r_S}{r} + \frac{r_Q^2}{r^2} \right)^{-1} dr^2$$
$$+ r^2 d\theta^2 + r^2 \sin^2 \theta \, d\phi^2 ,$$

where $r_S = 2GM/c^2$ is the Schwarzschild radius, and r_Q is the charge radius, given by

$$r_Q^2 = \frac{Q^2 G}{4\pi \varepsilon_0 c^4} .$$

There is also a radial electric field $E_r = Q/4\pi\varepsilon_0 r^2$. This is known as the Reissner–Nordström solution.

a) Identify the values of r for which the metric has singularities, in the case $r_Q < r_S/2$.

b) Solve for the radial geodesics outside the outermost singularity. Is there an event horizon there?

c) An extreme Reissner–Nordström black hole has $r_Q = r_S/2$. What happens to the singularities in this case? Why might we distrust the solution when $r_Q > r_S/2$?

d) Calculate $R^{tr}{}_{tr}$ for this metric, and comment on the metric singularities.

e) Calculate also $R_{\theta\theta}$. Why doesn't it vanish?

11.5* Consider a spaceship which is firing its thrusters in order to maintain a fixed position with $r_0 = 2r_S$ outside a black hole with Schwarzschild radius 10 light seconds.

a) Its crew members send out laser pulses which, according to them, are sent one second apart. They are received by a second ship located at a much larger fixed radius. What is the time interval between the pulses measured by the distant spaceship?

b) The spaceship at r_0 runs out of power and goes into free-fall, continuing to emit pulses. How many pulses escape the black hole, and how many pulses are emitted before the spaceship encounters the central singularity?

11.6* Recall Problem 9.4, where you studied circular orbits in a Schwarzschild space–time.

a) What is the smallest-possible specific angular momentum for a circular orbit? Calculate the orbital radius at this minimum specific angular momentum. This is known as the *innermost stable circular orbit* (ISCO).

b) What is the orbital period as measured by (i) an astronaut in this orbit and (ii) a distant observer?

11.7 This question is about the orbits of photons in the Schwarzschild space–time.

a) Verify that there is a circular orbit for photons at a radius $r_\circ = 3r_S/2$.

b) For a photon orbiting just outside $r_\circ$ at radius $r = r_\circ + \delta r$, show that the departure from the circular orbit δr has growing solutions which are exponential in the affine parameter. If δr is exponentially growing, by how much does it increase after one orbit?

11.8 The black-hole entropy is proportional to the area of the event horizon. Two black holes, whose masses are 8 Solar masses and 6 Solar masses respectively, undergo a merger. What is the minimum mass of the resulting black hole? Give an argument that the merger of two black holes is an irreversible process.

11.9 What is the wavelength of the peak emission of Hawking radiation from a black hole? How does this compare to the Schwarzschild radius?

11.10* A blackbody of unit area emits power according to the Stefan–Boltzmann law

$$P = \frac{\pi^2 k_B^4}{60c^2\hbar^3} T^4 .$$

Compute the power emitted via Hawking radiation by a black hole of mass M, and use this to obtain a differential equation for the evolution of the mass of an evaporating black hole. Solve this for a black hole of initial mass M_0 to show that the black hole lifetime is

$$\tau = 5120\pi \frac{G^2}{c^4\hbar} M_0^3 \simeq (8.4 \times 10^{-17}\,\mathrm{s\,kg^{-3}}) \times M_0^3 \,.$$

What is the predicted lifetime of a black hole whose initial mass is 10 Solar masses? What initial mass would give a lifetime equal to the age of the Universe (approximately 14 billion years)?

[Note: the Stefan–Boltzmann law gives the emission of electromagnetic radiation only, while a black hole may be able to radiate additional particle species if their mass–energy is below $k_B T$. If evaporation is into g_* particle degrees of freedom, the above formula is multiplied by $g_*/2$, allowing for the two photon polarisation states. A sophisticated calculation would allow g_* to vary with temperature as the black hole heats up during evaporation.]

Chapter 12

Cosmology

General Relativity lets us describe the entire Universe · first we find
the right metric, and then the equations for how it evolves

Einstein's theory enables us to address the grandest setting of all, the evolution of the entire Universe. The first rigorous cosmological models were developed by Willem de Sitter and Einstein shortly after General Relativity was finalised. At the time, it was not known that the Universe was expanding and so Einstein sought static models of the Universe, which necessitated inclusion of the cosmological constant. In the mid-1920s, Alexander Friedmann found a set of expanding Universe models, in absence of a cosmological constant, which have formed the basis of subsequent cosmological modelling. At the end of that decade, Edwin Hubble discovered the expansion of the Universe, beginning a sequence of discoveries that would establish the Hot Big Bang model as the standard description of the cosmos. As a final twist, at the end of the 20th century, it became accepted that the present Universe is not just expanding, but accelerating, leading to the reintroduction of the cosmological constant.

In this chapter and the next, our aim is to provide a brief yet rigorous derivation of simple cosmological models, so that the reader can go on to a study of physical and astrophysical aspects of cosmology with a clear understanding of how the underlying dynamics of the Universe follow from the principles of General Relativity.

Other sections of this book have focused on vacuum solutions to the Einstein equations, describing the space–time outside stars or black holes, or in the presence of gravitational waves. Cosmology provides the simplest context where we need to include matter.

We begin by assuming the **cosmological principle**, which states that all observers are equivalent, not just in terms of the laws of physics but in their environment at a given time too. Our description of the Universe should therefore not

Introducing General Relativity, First Edition. Mark Hindmarsh and Andrew Liddle.
© 2022 John Wiley & Sons Ltd. Published 2022 by John Wiley & Sons Ltd.
Companion website: www.wiley.com/go/hindmarsh/introducingGR

depend either on position or on the direction in which we are looking. These two properties are referred to as **homogeneity** and **isotropy**. They are independent requirements. For example, a Universe pervaded by a uniform magnetic field is homogeneous but has a preferred direction, while any spherically symmetric matter distribution around us would be isotropic but not necessarily homogenous. Adopting these assumptions is obviously an idealisation, as the real Universe contains plentiful structures such as stars and galaxies, but we can hope that it applies on average provided we consider large enough scales, much larger than the typical separation between galaxies. Observations indeed support this hypothesis in the present Universe, the most decisive being the direct observation of microwave radiation — the **cosmic microwave background (CMB)** — coming from all directions at almost uniform intensity. We will later interpret the CMB as remnant radiation left over from an initial hot phase of the Universe's evolution.

Our first task is to find what possible spatial geometries are consistent with homogeneity and isotropy.

12.1   Constant-curvature spaces

To be homogeneous, a space needs to have the same curvature at each point. Such spaces are known as constant-curvature spaces. From isotropy, we can choose a spherical coordinate system for our space, so

$$ds_3^2 = A(r)dr^2 + r^2 \left(d\theta^2 + \sin^2\theta \, d\phi^2\right), \tag{12.1}$$

where ds_3^2 is just the spatial part of the metric. This follows from the same line of argument used to establish the general form of the Schwarzschild space–time at the start of Chapter 8.

We can immediately save ourselves considerable effort by realizing that the above form is exactly the same as the general static spherical space–time, equation (8.1), except that $B(r)$ is set to zero, and there is no longer a time coordinate at all. So we can copy across the Riemann tensor elements from Chapter 8, setting $B(r)$ equal to zero:

$$R^{r\theta}_{\ r\theta} = R^{r\phi}_{\ r\phi} = \frac{1}{2r}\frac{A'}{A^2} ; \quad R^{\theta\phi}_{\ \theta\phi} = \frac{1}{r^2 A}(A-1). \tag{12.2}$$

The homogeneity assumption demands that the curvature of the space be independent of r, and in particular that the Ricci scalar should be a constant. Anticipating the final result for the metric, we will call that constant $6k$, so that

$$^{(3)}R = 2\left(R^{r\theta}_{\ r\theta} + R^{r\phi}_{\ r\phi} + R^{\theta\phi}_{\ \theta\phi}\right) = 6k, \tag{12.3}$$

where the factor 2 arises from the pair symmetry on the indices and the superscript in $^{(3)}R$ reminds us that this is the three-dimensional Ricci scalar. Hence,

$$\frac{1}{r}\frac{A'}{A^2} + \frac{1}{r^2}\left(1 - \frac{1}{A}\right) = 3k. \tag{12.4}$$

This equation has the general solution $A^{-1} = 1 - kr^2 - C/r$, where C is an integration constant. However, additionally demanding that the space becomes locally flat as $r \to 0$ fixes $C = 0$ (for more detail on this see Problems 12.1 — 12.3).

The function A is therefore

$$A(r) = \frac{1}{1 - kr^2}, \tag{12.5}$$

and so

$$ds_3^2 = \frac{dr^2}{1 - kr^2} + r^2\left(d\theta^2 + \sin^2\theta \, d\phi^2\right), \tag{12.6}$$

is the most general spatial metric with constant curvature. Note that the components of the Riemann tensor in its rank-$\binom{2}{2}$ form are also constant, and equal to each other

$$R^{r\theta}_{\ \ r\theta} = R^{r\phi}_{\ \ r\phi} = R^{\theta\phi}_{\ \ \theta\phi} = k. \tag{12.7}$$

This comes about because the rank-$\binom{2}{2}$ elements with identical index pairs contain the sectional curvature of the two-dimensional subspaces defined by those index pairs, as introduced in Section 5.8. In a homogeneous and isotropic space, the sectional curvatures should be constant and equal.

To interpret the metric we have found, it is helpful to change coordinates.

$k = 0$: This just gives flat space written in spherical coordinates. If we wish, we can change to Cartesian coordinates to write $ds_3^2 = dx^2 + dy^2 + dz^2$.

$k > 0$: There apparently is a singularity at $r = 1/\sqrt{k}$, but in fact it is a coordinate singularity. Introduce a new coordinate χ obeying $\sin\chi = \sqrt{k}\,r$. The metric becomes

$$ds_3^2 = \frac{1}{k}\left[d\chi^2 + \sin^2\chi\left(d\theta^2 + \sin^2\theta \, d\phi^2\right)\right], \tag{12.8}$$

which is the metric for a three-sphere of radius $1/\sqrt{k}$. A universe with $k > 0$ is referred to as a **closed universe**, in recognition of its finite volume.

$k < 0$: This situation is much harder to visualise, but putting $\sinh\chi = \sqrt{-k}\,r$

shows we have a hyperbolic three-space (sometimes also called Lobachev-skian after one of its discoverers). A universe with $k < 0$ is called an **open universe**.

The constant k measures the curvature of space, and so has dimensions of inverse length squared. If we wish, we can define a new dimensionless radial co-ordinate $\tilde{r} = \sqrt{|k|}\, r$, so that k takes on one of the values -1, 0, and $+1$. However, there is no transformation which can switch us amongst these three choices. Employing $\tilde{r}$ means that the radius is being measured in units of the curvature radius, and this is quite standard practice amongst relativists. Astrophysical cosmologists prefer to use physical units for distance, such as megaparsecs, in which case k cannot be rescaled in that way.

12.2   The metric of the Universe

Having obtained our constant curvature space, we need to incorporate it into a space–time. By assumption, the only further dependences we can put in are time dependences; in particular we can allow our space to grow or shrink with time. This leads us to the **Friedmann–Lemaître–Robertson–Walker (FLRW) metric**

$$ds^2 = -c^2 dt^2 + a^2(t) \left[\frac{dr^2}{1 - kr^2} + r^2 \left(d\theta^2 + \sin^2 \theta \, d\phi^2 \right) \right]. \qquad (12.9)$$

It looks at first that there could also have been a function of time, e.g. $b^2(t)$, before the dt^2, similarly to the Schwarzschild case. But such a function could immediately be removed by introducing a new time coordinate $dt' = b(t)dt$ and so the above expression is already the most general.

The quantity $a(t)$ is known as the scale factor of the Universe, and measures how the distance between points at fixed spatial coordinates changes with time. This metric was first used by Friedmann and then rediscovered by Georges Lemaître, both in the 1920s. Later, in the 1930s, Howard Robertson and Arthur Walker showed that its uniqueness follows entirely from geometry, without reference to the field equations of General Relativity, as indeed, we have seen in our derivation. It is often referred to not only as the FLRW metric but also as just FRW or RW. The role of the Einstein equations is to provide the specific form of $a(t)$ corresponding to a given energy–momentum tensor.

12.3   The matter content of the Universe

We consider a simple model in which the contents of the Universe are described as a single perfect fluid, using the formalism of Chapter 6 and in particular equation

(6.26). Chapter 13 will generalise to the realistic case where there are several separate components making up the Universe's constituents. By homogeneity and isotropy, the fluid's properties can depend only on time, and the 4-velocity u^μ cannot have any spatial components as they would break isotropy by selecting a special direction. Hence, $u^0 = c$ and $u^i = 0$, and we have

$$T^0_{\ 0} = -c^2 \rho(t); \quad T^i_{\ j} = \delta^i_{\ j} p(t), \tag{12.10}$$

where ρ is the mass density and p the pressure.

12.4   The Einstein equations

As with the Schwarzschild space–time, we are going to calculate the field equations stage-by-stage. We start with the Christoffel symbols, using the formula (5.12). After taking into account symmetry under exchange of the two lower indices, there are 13 non-zero terms in all. The angular ones we can take already from the toy example of the two-sphere, equation (5.34), as the scale factor cancels out in the expression:

$$\Gamma^\theta_{\ \phi\phi} = -\sin\theta\,\cos\theta\ ; \quad \Gamma^\phi_{\ \theta\phi} = \frac{\cos\theta}{\sin\theta}\,. \tag{12.11}$$

We write out the remaining non-zero components, using dots for time derivatives:

$$\Gamma^r_{\ tr} = \Gamma^\theta_{\ t\theta} = \Gamma^\phi_{\ t\phi} = \frac{\dot{a}}{a}; \quad \Gamma^\theta_{\ r\theta} = \Gamma^\phi_{\ r\phi} = \frac{1}{r}; \tag{12.12}$$

$$\Gamma^t_{\ rr} = \frac{1}{c^2}\frac{a\dot{a}}{1-kr^2}; \quad \Gamma^t_{\ \theta\theta} = \frac{a\dot{a}r^2}{c^2}; \quad \Gamma^t_{\ \phi\phi} = \frac{a\dot{a}r^2\sin^2\theta}{c^2};$$

$$\Gamma^r_{\ rr} = \frac{kr}{1-kr^2}; \quad \Gamma^r_{\ \theta\theta} = -r(1-kr^2); \quad \Gamma^r_{\ \phi\phi} = -r\sin^2\theta\,(1-kr^2).$$

Not all of these are actually needed in what follows.

From the Christoffel symbols, we obtain the Ricci tensor from its definition

$$R_{\nu\beta} = \Gamma^\mu_{\ \beta\nu,\mu} - \Gamma^\mu_{\ \mu\nu,\beta} + \Gamma^\mu_{\ \mu\rho}\,\Gamma^\rho_{\ \beta\nu} - \Gamma^\mu_{\ \beta\rho}\,\Gamma^\rho_{\ \mu\nu}\,. \tag{12.13}$$

The four diagonal terms are the only non-zero ones, the simplest being

$$R_{tt} = -3\frac{\ddot{a}}{a}\,. \tag{12.14}$$

They are mostly conveniently expressed in the form with one raised index, which makes the three spatial terms identical. The result is

$$R^t{}_t = \frac{3}{c^2}\frac{\ddot{a}}{a}; \quad R^r{}_r = R^\theta{}_\theta = R^\phi{}_\phi = \frac{1}{c^2}\frac{\ddot{a}}{a} + \frac{1}{c^2}\frac{2\dot{a}^2 + 2kc^2}{a^2}, \quad (12.15)$$

Finally, this leads to the Ricci scalar

$$R = R^\mu{}_\mu = \frac{6}{c^2}\left(\frac{\ddot{a}}{a} + \frac{\dot{a}^2 + kc^2}{a^2}\right). \quad (12.16)$$

All of the above calculations are purely about the geometry of the space–time. We see that the quantities describing its curvature involve time derivatives up to second order. The Einstein equation relates them to the matter content, and so gives a dynamical equation describing the evolution of the scale factor a.

For simplicity, for now, we will suppose that any cosmological constant Λ is included in the energy–momentum tensor, and use its equivalent density and pressure (7.12) to reintroduce it later. The t–t Einstein equation is

$$R^t{}_t - \frac{1}{2}R = \frac{8\pi G}{c^4}T^t{}_t, \quad (12.17)$$

which gives

$$\left(\frac{\dot{a}}{a}\right)^2 + \frac{kc^2}{a^2} = \frac{8\pi G}{3}\rho. \quad (12.18)$$

This is known as the **Friedmann equation** and describes how the scale factor of the Universe evolves in response to the presence of a matter density $\rho(t)$ and spatial curvature k.

The r–r Einstein equation gives a second equation; a similar calculation (see Problem 12.4) yields

$$2\frac{\ddot{a}}{a} + \left(\frac{\dot{a}}{a}\right)^2 + \frac{kc^2}{a^2} = -8\pi G\frac{p}{c^2}. \quad (12.19)$$

But an alternative is to instead use energy–momentum conservation, considering $T^\mu{}_{t;\mu} = 0$.[1] Writing this in full gives

$$T^\mu{}_{t,\mu} + \Gamma^\mu{}_{\alpha\mu}T^\alpha{}_t - \Gamma^\alpha{}_{t\mu}T^\mu{}_\alpha = 0. \quad (12.20)$$

---

[1]The momentum conservation equations $T^\mu{}_{i;\mu} = 0$ do not give any additional information as isotropy ensures the momentum flow is identically zero and hence automatically conserved.

Using the Christoffel symbols given above, we get

$$-c^2\dot{\rho} - 3\frac{\dot{a}}{a}\rho c^2 - 3\frac{\dot{a}}{a}p = 0. \tag{12.21}$$

We rearrange to find

$$\dot{\rho} + 3\frac{\dot{a}}{a}\left(\rho + \frac{p}{c^2}\right) = 0, \tag{12.22}$$

which is known as the **fluid equation**, describing the evolution of the density ρ in response to the expansion of the Universe and its pressure. Notice that it does not depend explicitly on the spatial curvature k, only indirectly via curvature's influence on a.

Subtracting the Friedmann equation from equation (12.19) yields the simpler form known as the **acceleration equation** (sometimes also called a Friedmann equation)

$$\frac{\ddot{a}}{a} = -\frac{4\pi G}{3}\left(\rho + \frac{3p}{c^2}\right). \tag{12.23}$$

We could also have derived this by differentiating the Friedmann equation (12.18) and eliminating $\dot{\rho}$ using the fluid equation, so the computation of the r–r Einstein equation is indeed unnecessary if we use the energy conservation equation.

Finally, we reintroduce the cosmological constant, by recalling the equivalent density and pressure of a fluid whose energy–momentum tensor reproduces the term $\Lambda g_{\mu\nu}$ in the Einstein equation (see equation (7.12)). The Friedmann and acceleration equations become

$$\left(\frac{\dot{a}}{a}\right)^2 + \frac{kc^2}{a^2} - \frac{\Lambda}{3} = \frac{8\pi G}{3}\rho\,; \tag{12.24}$$

$$\frac{\ddot{a}}{a} - \frac{\Lambda}{3} = -\frac{4\pi G}{3}\left(\rho + \frac{3p}{c^2}\right). \tag{12.25}$$

The fluid representing the cosmological constant separately obeys the fluid equation (12.22), and so the fluid equation is also satisfied by the other contents of the Universe.

One of the simplest cosmological solutions is a Universe with a cosmological constant but no other contents, i.e. $\rho = p = 0$. This model is known as **de Sitter space** and you can explore it via Problems 12.6 and 12.7. It has the curious property that depending on the choice of coordinate system, the solution can look like a rapidly expanding cosmology or like a static spherically symmetric space–time.

Chapter 13 explores how these equations can be applied to understand the evolution of our Universe.

Problems

12.1* Show by explicit calculation that the general solution of equation (12.4) is indeed

$$A^{-1} = 1 - kr^2 - \frac{C}{r},$$

where C is an integration constant. By considering the behaviour of the Riemann tensor components as $r \to 0$, show that the requirement of local flatness demands $C = 0$.

12.2 Compute the sectional curvatures for the three-dimensional constant-curvature spaces of the previous problem, keeping the integration constant C in the calculation. What condition is implied by equality of the sectional curvatures?

12.3 [Hard!] Following on from Problem 12.1, a rigorous demonstration that local flatness fails at the origin when $C \neq 0$ requires evaluation of an invariant such as the Kretschmann scalar

$$K \equiv R_{ijkl}R^{ijkl},$$

defined on page 141. Compute this to demonstrate that K diverges at the origin unless $C = 0$.
[Hint: $R^{r\theta}{}_{r\phi}$, $R^{r\theta}{}_{\theta\phi}$, and $R^{r\phi}{}_{\theta\phi}$ are all zero.]

12.4 Confirm by explicit calculation the expressions for the four non-zero Ricci tensor elements in the FLRW metric, as given in equation (12.15). Verify that the r–r Einstein equation for a homogeneous and isotropic Universe is (in absence of a cosmological constant) given by equation (12.19).

12.5* The equation for a three-sphere of radius ρ embedded in a four-dimensional Euclidean space is $x^A x^A = \rho^2$, where $A = 1, 2, 3, 4$. By making a transformation to four-dimensional spherical polar coordinates ρ, χ, θ, ϕ defined by

$$
\begin{aligned}
x^1 &= \rho \sin \chi \sin \theta \cos \phi; \\
x^2 &= \rho \sin \chi \sin \theta \sin \phi; \\
x^3 &= \rho \sin \chi \cos \theta; \\
x^4 &= \rho \cos \chi,
\end{aligned}
$$

show that the line element dl on surfaces of constant ρ is

$$
dl^2 = \rho^2 d\chi^2 + \rho^2 \sin^2 \chi (d\theta^2 + \sin^2 \theta \, d\phi^2).
$$

12.6* Consider a flat vacuum Universe with a positive cosmological constant. Write down the Friedmann equation in such a universe, and find its general solution $a(t)$. This space–time is known as de Sitter space and can be written in a variety of different coordinate systems.

12.7 Evaluate the Ricci scalar for the de Sitter space solution found in the previous question and comment on your result.

Chapter 13

Cosmological Models

*To make cosmological models, we solve the Friedmann equation ·
how does light travel in an expanding Universe? · what makes a
realistic model of our own Universe?*

In Chapter 12, we derived the equations that describe the evolution of a homogeneous and isotropic Universe. They are the Friedmann equation

$$\left(\frac{\dot{a}}{a}\right)^2 = \frac{8\pi G}{3}\rho - \frac{kc^2}{a^2} + \frac{\Lambda}{3}, \tag{13.1}$$

whose solution determines the expansion history $a(t)$, and the fluid equation

$$\dot{\rho} + 3\frac{\dot{a}}{a}\left(\rho + \frac{p}{c^2}\right) = 0, \tag{13.2}$$

which gives the evolution of the density. The solutions will depend on the values
of the two constants, the spatial curvature k and cosmological constant Λ, as well
as the relation between the pressure p and density ρ of the Universe's contents.
This last relation is known as the **equation of state**. For the time being we are
considering the Universe as containing a single perfect fluid, but a full description
will need to consider several coexisting fluids each with their own fluid equation
and with possible energy exchange amongst them.

13.1 Simple solutions: matter and radiation

The microphysical properties of materials are responsible for setting the relationship between pressure and density. There are two main limits of cosmological
interest; recapping from Section 6.2.5 we have

Introducing General Relativity, First Edition. Mark Hindmarsh and Andrew Liddle.
© 2022 John Wiley & Sons Ltd. Published 2022 by John Wiley & Sons Ltd.
Companion website: www.wiley.com/go/hindmarsh/introducingGR

1. **'Matter' (sometimes called 'dust'):** This applies to particles which move slowly compared to the speed of light. The fluid might be comprised of individual fundamental particles, or might be a fluid made of composite objects such as stars, galaxies, or even black holes. The pressure of a cosmological matter fluid is zero, which is a good approximation for these cases.

2. **'Radiation':** This applies to particles moving at ultra-relativistic velocities (meaning $\gamma \gg 1$). The simplest example is a gas of photons, which by definition are moving at the speed of light. Alternatively, we may have particles whose rest mass is negligible compared to their kinetic energy, which is thought to be the case for neutrinos for most of cosmic history and may well have applied to most or all known particle species in the hot early stages of the Big Bang cosmology that we will be describing. A radiation fluid in a cosmological setting possesses a pressure $p = \rho c^2/3$, which is an increasingly good approximately as the average speed of the particles approaches that of light. If the particles are close to thermal equilibrium, the density of a radiation fluid is proportional to its temperature to the fourth power.

We refer to these two possibilities by attaching a subscript 'm' or 'r' to the density ρ.

In each case the fluid equation is readily solved. For matter we have

$$\dot{\rho}_m = -3\frac{\dot{a}}{a}\rho_m .$$

(13.3)

This is a separable equation whose solution is

$$\rho_m = \frac{\rho_{m,0}}{a^3} ,$$

(13.4)

where the boundary condition is that $\rho_m = \rho_{m,0}$ when $a = 1$. This shows that the density falls in proportion to the increasing volume of the Universe, which is proportional to a^3. We will always use the subscript '0' to refer to quantities evaluated at the present time t_0.

Similarly, for radiation we have

$$\dot{\rho}_r = -4\frac{\dot{a}}{a}\rho_r ,$$

(13.5)

with solution

$$\rho_r = \frac{\rho_{r,0}}{a^4} .$$

(13.6)

Here the density falls by an additional factor of a over and above the volume, which we will see is attributed to a loss of energy — the gravitational **redshifting** — of individual particles. If the radiation is in a thermal state at temperature T, this corresponds to a cooling $T \propto 1/a$.

In each case, the solution holds regardless of the values of k and Λ as we did not need to use the Friedmann equation.

To find the corresponding expansion rates, we now need to solve the Friedmann equation. For simplicity we here consider the case $k = \Lambda = 0$. For a Universe containing only matter, the Friedmann equation is

$$\left(\frac{\dot{a}}{a}\right)^2 = \frac{8\pi G}{3}\frac{\rho_{m,0}}{a^3}. \tag{13.7}$$

Once again the equation is separable,

$$a^{1/2}da = \frac{8\pi G\rho_{m,0}}{3}dt, \tag{13.8}$$

leading to the solution

$$a(t) = \left(\frac{t}{t_0}\right)^{2/3}. \tag{13.9}$$

The scale factor is determined only up to a multiplicative constant of integration, and we follow the common convention that a is normalised to unity at the present time t_0.

A similar calculation for a Universe containing only radiation gives

$$a(t) = \left(\frac{t}{t_0}\right)^{1/2}. \tag{13.10}$$

In each case, the density scales as $1/t^2$.

Our actual Universe contains both matter and radiation, which presently do not exchange significant amounts of energy, and so the total density appearing in the Friedmann equation is comprised of two parts, $\rho = \rho_m + \rho_r$, with each component obeying its own fluid equation. Those fluid equations will continue to have the evolutions given by equations (13.4) and (13.5), and Problem 13.3 shows how the combined system can be solved. As we will describe later, in the present Universe the density of matter dominates over that of radiation by a factor of a few thousand (and even the former is dominated by the equivalent energy density of the cosmological constant). However, if we track back in time, the stronger evolution of the radiation means that it dominated at early epochs, explaining the terminology of the **Hot Big Bang**. The early Universe was in a hot dense radiation-dominated phase, expanding as $a \propto t^{1/2}$ until such time as matter and radiation reached equal densities. Thereafter matter domination took over, the scale factor evolution switching smoothly to $a \propto t^{2/3}$.

During the early stages, the photons had enough energy to maintain the matter as an ionised plasma, and the interactions between the photons and the charged particles were rapid enough to keep them very close to thermal equilibrium. Subsequent cooling due to the expansion, $T \propto 1/a$, caused the individual photons to lack the energy to prevent electrons binding to nuclei to form atoms. After this time the photons could no longer interact with them and travelled freely through the Universe. This process is known as **decoupling**, and took place when the Universe was around $1/1100$ of its present size.[1] This cooled radiation is readily observed today, and is known as the cosmic microwave background (CMB) due to its current characteristic frequency lying in that part of the electromagnetic spectrum. Gravitational redshifting preserves the shape of the thermal distribution, which by now has cooled to an observed temperature of $T_0 \simeq 2.725 \, \text{K}$.

In many cosmological solutions, including those above, we find that the scale factor $a(t)$ was zero at some initial finite time t. At this time the metric breaks down, the density and curvature being infinite at that instant. Note that t also measures the elapsed proper time for comoving observers, which is hence also finite. While it was originally thought that this initial singularity might be an artefact of the assumed high degree of symmetry of the FLRW metric, Hawking found a cosmological analogue to Penrose's black-hole singularity theorem to show that a singularity was inevitable under a broad range of circumstances. The initial singularity would correspond to the actual Big Bang (which however occurs on a space-like surface and so should not be thought of as being at a single space–time point).

It is however notable that one of the theorem's assumptions is the strong energy condition $\rho + 3p > 0$, which is violated if the very early Universe underwent a period of accelerated expansion, dubbed **inflation**. Inflation is indeed widely believed to have taken place — see Section 13.4. Attempts to extend Hawking's theorem to include inflationary cosmologies are not yet completely watertight, and indeed in some versions of open universe cosmologies the singularity can be shown to be a coordinate one (corresponding to the surfaces of constant t rotating from space-like to null). Additionally, the singularity theorems are results of classical General Relativity and almost all physicists expect that quantum effects are important in the vicinity of singularities, making their true nature and even existence highly uncertain.

---

[1] A detailed treatment distinguishes 'recombination', when the majority of electrons are incorporated into atoms, from the slightly later phenomenon of 'decoupling' marking the time after which the majority of photons will never again interact.

13.2   Light travel, distances, and horizons

Returning to the general case with spatial curvature and a cosmological constant, we now study some kinematical properties of the Universe and how light travels within it.

13.2.1   Light travel in the cosmological metric

The metric is equation (12.9), repeated here for convenience:

$$ds^2 = -c^2 dt^2 + a^2(t) \left[\frac{dr^2}{1 - kr^2} + r^2 \left(d\theta^2 + \sin^2\theta \, d\phi^2 \right) \right] . \qquad (13.11)$$

Light follows null rays $ds = 0$, so radial rays (i.e. $d\theta = d\phi = 0$) satisfy

$$\int \frac{dr}{\sqrt{1 - kr^2}} = \int \frac{cdt}{a(t)} . \qquad (13.12)$$

A simpler version of the above equation can be obtained by changing the radial variable so that the curvature term in the metric shifts onto the angular degrees of freedom, via $\sqrt{k}\,\tilde{r} = \sin^{-1}(\sqrt{k}\,r)$ in the closed case and $\sqrt{-k}\,\tilde{r} = \sinh^{-1}(\sqrt{-k}\,r)$ in the open case. Then the above equation becomes

$$\int d\tilde{r} = \int \frac{cdt}{a(t)} , \qquad (13.13)$$

regardless of spatial geometry. The corresponding metric is

$$ds^2 = -c^2 dt^2 + a^2(t) \left[d\tilde{r}^2 + S^2(\tilde{r}) \left(d\theta^2 + \sin^2\theta \, d\phi^2 \right) \right] , \qquad (13.14)$$

where $S(\tilde{r})$ equals $\tilde{r}$ in the flat case, equals $\sin(\sqrt{k}\,\tilde{r})/\sqrt{k}$ in the closed case $k > 0$, and equals $\sinh(\sqrt{-k}\,\tilde{r})/\sqrt{-k}$ in the open case $k < 0$.

Another form of the metric that can be useful is to change the time variable to **conformal time**, defined by $d\tau = cdt/a(t)$, which makes the scale factor an overall multiplier, for instance the original form becoming

$$ds^2 = a^2(\tau) \left[-d\tau^2 + \frac{dr^2}{1 - kr^2} + r^2 \left(d\theta^2 + \sin^2\theta \, d\phi^2 \right) \right] . \qquad (13.15)$$

Along a light trajectory from the origin we find $\tau = \tilde{r}$, and hence conformal time is literally the distance that a light ray can have travelled since the Big Bang. Note that this is the distance as measured by the coordinate $\tilde{r}$, not the actual physical distance given by ds. The former is also known as the **comoving distance**, since the FLRW spatial coordinates are carried along with the expansion by the scale factor.

13.2.2   Cosmological redshift

In an expanding Universe, light changes frequency as it propagates due to gravitational redshift. To evaluate this, consider two consecutive pulses sent across a fixed comoving distance $\tilde{r}$, separated by an emission time dt_{emit} and an observation time dt_{obs}. For the first

$$\int_{t_{\text{emit}}}^{t_{\text{obs}}} \frac{c\,dt}{a(t)} = \int_0^{\tilde{r}_0} d\tilde{r}, \tag{13.16}$$

and for the second

$$\int_{t_{\text{emit}}+dt_{\text{emit}}}^{t_{\text{obs}}+dt_{\text{obs}}} \frac{c\,dt}{a(t)} = \int_0^{\tilde{r}_0} d\tilde{r}. \tag{13.17}$$

The right-hand sides are equal, so we can combine these equations. Subtracting off the common piece of the integrals from $t_{\text{emit}} + dt_{\text{emit}}$ to t_{obs} from each side leaves

$$\int_{t_{\text{emit}}}^{t_{\text{emit}}+dt_{\text{emit}}} \frac{c\,dt}{a(t)} = \int_{t_{\text{obs}}}^{t_{\text{obs}}+dt_{\text{obs}}} \frac{c\,dt}{a(t)}. \tag{13.18}$$

Taking the differences to be infinitesimal then gives

$$\frac{c\,dt_{\text{obs}}}{a(t_{\text{obs}})} = \frac{c\,dt_{\text{emit}}}{a(t_{\text{emit}})}. \tag{13.19}$$

The interval between pulses has changed in proportion to the change of scale factor between emission and reception.

Considering the pulses to be consecutive crests of a light wave shows that its frequency ν changes according to

$$\frac{\nu_{\text{emit}}}{\nu_{\text{obs}}} \equiv 1 + z = \frac{a(t_{\text{obs}})}{a(t_{\text{emit}})}, \tag{13.20}$$

which defines the **redshift** z, with t_{obs} usually taken as the present epoch. If the Universe is expanding, the frequency at observation is reduced compared to that at emission, and we say that the light has been redshifted. This is the microphysical origin of the additional factor $1/a$ in the evolution law (13.6) for the density of radiation as compared to matter.

Measurement of the position of narrow spectral lines is a powerful and direct measure of the epoch of emission of radiation from distant objects. An object, such as a distant galaxy or supernova, is said to be at redshift z if its distance is such that its light is observed by us to have that redshift. The redshift will be greater than zero in an expanding Universe, with $z \to \infty$ as we consider light emitted ever closer to the Big Bang itself.

It is sometimes loosely stated that the cosmological redshift is a Doppler effect due to the recession velocity of distant galaxies. The form of the cosmological redshift in equation (13.20) shows that it is really due to the expansion of space, although the Doppler formula does give the correct numerical value in the limit of redshifts much less than 1 (see Problem 13.5).

13.2.3 The expansion rate

The expansion rate of the Universe is given by the **Hubble parameter**, defined as

$$ H \equiv \frac{\dot{a}}{a}. \tag{13.21} $$

Its present value is called the **Hubble constant**, denoted H_0, which is one of the fundamental cosmological parameters that needs to be determined from observations.

We rewrite the Friedmann equation by first defining the **critical density**, which is the density the Universe would need to have, for a given expansion rate, to give a flat Universe:

$$ \rho_{\text{cr}} \equiv \frac{3H^2}{8\pi G}. \tag{13.22} $$

We then define all densities relative to this using the **density parameters**

$$ \Omega_i = \frac{\rho_i}{\rho_{\text{cr}}}, \tag{13.23} $$

for each material with density ρ_i, and we normalise the scale factor so that its present value a_0 equals 1. The Friedmann equation can then be written using the scale factor as the independent variable, in the form

$$ H^2(a) \equiv \left(\frac{\dot{a}}{a}\right)^2 = H_0^2 \left[\Omega_{\Lambda,0} + \Omega_{\text{m},0}\, a^{-3} + \Omega_{\text{r},0}\, a^{-4} - (\Omega_0 - 1)\, a^{-2} \right], \tag{13.24} $$

where the density parameters are their *present* values, corresponding to the cosmological constant ($\Omega_{\Lambda,0} \equiv \Lambda/3H^2$), non-relativistic matter $\Omega_{m,0}$, and relativistic matter $\Omega_{r,0}$, while Ω_0 is the sum of these and equals one in the case of a flat Universe.

13.2.4 The age of the Universe

The above equation is separable

$$H_0 dt = \frac{da}{a\sqrt{\Omega_{\Lambda,0} + \Omega_{m,0} a^{-3} + \Omega_{r,0} a^{-4} - (\Omega_0 - 1) a^{-2}}}, \quad (13.25)$$

and can be integrated to give the age as a function of scale factor, $t(a)$. However only in some cases can the integral be performed analytically, and even then it typically does not yield a function which can be analytically inverted to give $a(t)$. We can obtain the current age of the Universe by integrating this up to scale factor of one.

As a specific example, consider the most observationally relevant case of a flat Universe with matter and a cosmological constant but negligible radiation, which simplifies this to

$$H_0 dt = \frac{da}{a\sqrt{1 - \Omega_{m,0} + \Omega_{m,0} a^{-3}}}. \quad (13.26)$$

Even then it is quite a bit of work to solve this, to find two equivalent forms (see Problem 13.6)

$$\begin{aligned} H_0 t_0 &= \frac{2}{3} \frac{1}{\sqrt{1 - \Omega_{m,0}}} \ln\left[\frac{1 + \sqrt{1 - \Omega_{m,0}}}{\sqrt{\Omega_{m,0}}}\right]; \\ &= \frac{2}{3} \frac{1}{\sqrt{1 - \Omega_{m,0}}} \sinh^{-1}\left[\sqrt{\frac{1 - \Omega_{m,0}}{\Omega_{m,0}}}\right]. \end{aligned} \quad (13.27)$$

The right-hand side is typically of order unity, meaning that the age of the Universe is approximately H_0^{-1}, known as the Hubble time.

13.2.5 The distance–redshift relation and Hubble's law

Taking $a(t)$ to be normalised to unity at present, and noting the useful relationship

$$\frac{dz}{dt} = -\frac{\dot{a}}{a^2} = -\frac{H}{a}, \quad (13.28)$$

the distance travelled by light between redshift z and the present, via equation (13.13), is

$$\tilde{r} = \int_0^z \frac{c\,dz}{H(z)} \,. \tag{13.29}$$

The choice of the radial variable $\tilde{r}$ simplifies the form of the equation in the non-flat case.

For nearby objects we can assume H does not change while their emitted light propagates to us, so that $\tilde{r} \simeq cz/H_0$. We then find that their apparent recession velocity v from us due to the expansion, as inferred from their redshift, is related to the physical distance $x \equiv a\tilde{r}$ by

$$v = H_0 x \,. \tag{13.30}$$

This linear law is named **Hubble's Law** after its discoverer. Despite its simple form, it has proven difficult to measure Hubble's constant, because the physical distance is not directly observable. Instead it must be estimated via some other property, for instance using objects of assumed fixed luminosity known as **standard candles** (cepheid variable stars being an example). Moreover, Hubble's Law is spoiled by irregularities in the density distribution, which lead to galaxies moving with so-called **peculiar velocities** relative to the uniform expansion, which must be modelled and removed. At larger distances the law needs to be modified to allow for the variation of $H(z)$, and also to include spatial curvature if it is non-zero.

The Hubble constant is often expressed using a dimensionless parameter h via

$$H_0 = 100h\,\mathrm{km\,s^{-1}\,Mpc^{-1}} \,, \tag{13.31}$$

where the distances are measured in megaparsecs. This form was chosen to make h of order unity at a time when its value was not well known, but it is now well determined as $h \simeq 0.7$. The units of H_0 are simply inverse time and it can also be written as

$$H_0 = \frac{h}{9.77 \times 10^9 \,\mathrm{yrs}} \,. \tag{13.32}$$

Hence, the age of the Universe is predicted to about 14 billion years.

13.2.6 Cosmic horizons

The **particle horizon** is the maximum distance from which light can have reached us, starting out at the Big Bang when $z = \infty$. In a spatially flat matter-dominated

Universe, the integral in equation (13.29) cannot be done analytically but is well approximated by

$$\tilde{r} = 2cH_0^{-1}\Omega_{m,0}^{-0.4}\,. \tag{13.33}$$

For electromagnetic radiation, the actual furthest distance light can have travelled unimpeded is given by the integral since decoupling at around $z_{dec} \simeq 1100$, since before that photons are interacting with the ionised matter and the Universe effectively opaque. This distance to the last-scattering surface is often considered the boundary of our **observable Universe**, though in principle both neutrinos and gravitational waves could have travelled from somewhat further.

The **cosmological event horizon** is the boundary of the region to which we can send a signal from our present space–time location, i.e. it delimits the region of the Universe we can potentially influence. Equivalently, it lies at the present location of the furthest objects we will be able to see into the infinite future assuming we maintain our present coordinate location. Since the infinite future corresponds to redshift $z = -1$, the distance to the event horizon is given by equation (13.29) with integration limits 0 and -1. If the Universe is accelerating, the integral will converge, meaning there is a limit to how far we will ever be able to see in the Universe. If instead the acceleration were to cease in the future, there may be no event horizon. If the Universe were to recollapse, the integral would need to be done over the remaining expanding segment of the Universe and then the entire collapsing epoch; depending on the Universe's properties it might or might not be possible for an observer to witness the entire volume before recollapse.

13.2.7   The luminosity and angular-diameter distances

As already remarked, the coordinate distance $\tilde{r}$ to a distant object is in practice not directly observable. Observable distances relate to the measurable properties of size or luminosity of such objects. The luminosity distance and the angular-diameter distance refer to the apparent distances of objects based, respectively, on their observed brightnesses and their angular sizes. They are the distances that objects *appear* to have based on their observed properties. We will just state, rather than derive, the results.

In a spatially flat Universe, we have the simple relations

$$d_{lum} = \tilde{r}(1+z); \quad d_{ang} = \frac{\tilde{r}}{1+z}\,. \tag{13.34}$$

The former indicates the effect of redshift on the apparent brightness of an object, while the latter indicates that an object *of fixed physical size* had a corresponding larger comoving size when the Universe was smaller. The relation $d_{lum}/d_{ang} =$

$(1+z)^2$, known as the **Etherington reciprocity relation**, depends on redshift alone. It is a consequence of photon number conservation and holds generally in any theory of gravity based on a metric.

In a curved Universe ($k \neq 0$) we must also allow for the focusing effect of the geometry. This is most easily done using the metric in the form given in equation (13.14), and leads to the $S(\tilde{r})$ factor multiplying each of these distances (and hence cancelling in the reciprocity relation).

Each of these distances is extensively used in observational cosmology. The luminosity distance is used to interpret observations of Type Ia supernovae via the luminosity distance versus redshift relation, which in 1998 provided the first compelling evidence that the Universe is presently accelerating. The angular-diameter distance is needed to interpret the angular scale of features in both the CMB and in the matter distribution (the latter commonly referred to as baryon acoustic oscillations). The CMB features, known as anisotropies and shown later in Figure 13.1, currently give the most precise constraints on cosmological models.

13.3   Ingredients for a realistic cosmological model

Our discussion thus far has highlighted the different kinds of cosmological models that are possible, but which describes our own Universe? That question can only be answered through observation, and modern cosmology has access to a wide range of probes that tell us about the properties of our Universe.

Our starting point is an inventory of the material constituents of the Universe, which determine the energy–momentum tensor. We need five ingredients.

Baryons: Baryons are the protons and neutrons which make up the stars, galaxies, and hot gas that are the visible features of our Universe. And which of course we ourselves are made. By cosmological convention electrons are also included in this descriptor, despite technically being leptons; anyway charge neutrality dictates one electron per proton and the tiny mass of electrons relative to protons renders their contribution to the density negligible.

Photons: Electromagnetic radiation pervades the Universe. By density, the dominant form is the CMB left over from the Universe's early stages. While individual photons of starlight are much more energetic, they are very much rarer. The CMB carries a record of the Universe's youth.

Neutrinos: A key part of the Standard Model of Particle Physics, neutrinos are predicted to fill the Universe with a cosmic neutrino background whose density is comparable to the CMB. Their weakly interacting nature makes this background impossible to detect directly with currently conceivable technology, but their presence is inferred through their collective gravitational effect both on the overall expansion rate and on the development of density irregularities.

Dark matter: For almost a century it has been recognised that baryons do not provide sufficient gravitational attraction either to hold galaxies together or to even form them in the first place. Something else is doing that, which we call dark matter. It is not known what it is, but its gravitational influence is vital if we are to understand a wide range of observations. Dark matter is widely believed to be in the form of individual fundamental particles, though other possibilities such as compact astrophysical objects remain open. If the particles move at sub-relativistic velocities, the dark matter is referred to as **cold dark matter** (CDM) and its density provides a complete description of its cosmological effects; this is the usual assumption and fits the observational data well. Some theorists have instead proposed modifications to gravity to account for the effects usually attributed to dark matter, but as yet there is no theory matching the explanatory power and simplicity of the dark matter hypothesis.

Dark energy: It is widely accepted that the present Universe is accelerating, and none of the above ingredients can achieve that so we need something extra, which is known as dark energy. Equation (12.23) immediately shows that the dark energy must have negative pressure, indeed negative enough to overcome the positive pressure of other materials to ensure that the total $\rho + 3p/c^2$ is negative.[2] The simplest example is a cosmological constant Λ, which has $p = -\rho c^2$ as seen in equation (7.12). This choice matches all existing observations and we will adopt it, though the literature on possible alternatives is vast.

A cosmological model comprising these ingredients is commonly called ΛCDM (or sometimes LCDM) after its two dominant components. It can also be called the **standard cosmological model**.

One of the main goals of observational cosmology is to figure out how much of each ingredient is needed, by measuring their corresponding density parameters. The objective is to determine which, out of the family of possible Big Bang universes, corresponds to our own.

While the list above has five densities, the photon density is accurately determined from the blackbody spectrum of the CMB at observed temperature $T_0 = 2.7255 \pm 0.0006$ K, and the neutrino density can be inferred from it using thermodynamics arguments applied at an early epoch when they were strongly interacting with each other. Also, the usual assumption of spatial flatness (well justified

---

[2]Dark energy is not a very helpful name; its key property is its large negative effective pressure, not its energy (energy anyway being synonymous with mass).

Figure 13.1 An all-sky map of the CMB temperature from the final 2018 data release by the *Planck* Satellite collaboration. The variations are approximately one part in 10^5 of the mean temperature. (Figure courtesy ESA/Planck Collaboration.)

by observations) allows one further density to be written in terms of the others to ensure the density parameters sum to unity. This leaves two parameters describing the densities to be determined from observations, usually taken to be the baryon and dark matter densities, alongside the Hubble constant giving the current expansion rate.

We have already seen how to model the homogeneous Universe and derive observable quantities such as the luminosity and angular-diameter distances. This is sufficient to analyse several types of observation, sometimes referred to as kinematical probes, a class which includes direct measurements of the Hubble constant and the magnitude–redshift relation of distant Type Ia supernova which, after some calibration, are believed to be excellent standard candles.

A complete description of the history of our Universe must include not just the evolution of an idealised homogeneous and isotropic Universe, but the departures from homogeneity that give all the observed structures, such as galaxies, stars, and ultimately ourselves. Modelling these inhomogeneities, commonly called **density perturbations**, is now an extremely well-developed part of modern cosmology and the most powerful tool for learning about our Universe's properties. The mathematics is beyond our scope but covered in many of the more advanced textbooks in our Bibliography. At present the best source of constraints is the patterns of irregularities in the CMB, particularly its temperature but also the polarisation of the observed radiation. First detected by the COsmic Background Explorer (COBE) satellite in 1992, these **anisotropies** have now been mapped in exquisite detail by the European Space Agency's *Planck* satellite mission which published its final results in 2018 — see Figure 13.1. They prove to be in excellent agreement with the predictions of the standard cosmological model and determine

Name	Symbol	Value
Baryon density	$\Omega_b h^2$	0.02237 ± 0.00015
Cold dark matter density	$\Omega_c h^2$	0.1200 ± 0.0012
Sound horizon	$100\,\theta_{MC}$	1.0409 ± 0.0003
Optical depth	τ	0.054 ± 0.007
Perturbation amplitude	$\ln(10^{10} A_s)$	3.044 ± 0.014
Spectral index	n_s	0.965 ± 0.004
Hubble constant	h	0.674 ± 0.005
Matter density	Ω_m	0.315 ± 0.007
Cosmological constant density	Ω_Λ	0.685 ± 0.007
Age of the Universe	t_0	$(13.797 \pm 0.023) \times 10^9$ yrs

Table 13.1 The cosmological parameter values of our Universe, as according to the *Planck* satellite 2018 data analysis. These results use the *Planck* CMB measurements alone, and can be marginally improved by combining with other data. The upper six are the parameters actually fit to the data (not all of which have been introduced in this book), while the lower set are derived from those six. An up-to-date resource for current preferred values is the reviews section of the biennial Review of Particle Physics at `http://pdg.lbl.gov`.

all its main parameters to high precision.

To make predictions for the CMB anisotropies, we need a model for the initial perturbations present when the universe was very young (which we mention in the following section arise as a prediction of the cosmological inflation hypothesis). A simple model postulates these to be of power-law form, requiring specification of an amplitude (usually denoted A_s) and a scale-dependence (usually called the spectral index and denoted n_s), and this fits the data well. One further phenomenological parameter known as the optical depth τ, accounting for scattering of a small fraction of the CMB photons from ionised electrons on their way to us, is also necessary. Modelling the CMB therefore adds three parameters to the three mentioned above, giving a six-parameter standard cosmological model to be assessed against the data.

The six parameters can be specified in different ways by combining them algebraically, but particular choices are favoured by the technical details of the data analysis. Results on any particular parameter of interest can then be derived from these. The main *Planck* collaboration analysis uses the parameters shown in Table 13.1; this multiplies the density parameters by h^2 and uses a more complex quantity (the sound horizon at the last-scattering surface, θ_{MC}, expressed using the angular-diameter distance) in place of the Hubble constant. The table shows the resulting constraints, initially for the six parameters directly fit to the data, and then those derived for some of the more fundamental physical parameters that we used to develop our model. The matter density is the sum of the CDM and baryon densities.

The table shows that all the parameters of the standard cosmological model are accurately determined, often at the percent level or better, which is a stunning achievement. Tests for deviations from the standard model — for example allowing spatial curvature, non-standard neutrino properties, or simple modifications to General Relativity — have seen no evidence of them and strongly limit them. For example spatial flatness has been observationally verified to well within one percent accuracy.

While the CMB anisotropies provide (at time of writing) by far the most powerful cosmological constraints, there are a vast number of other types of observations available. These lend strong further support to the standard cosmology and help constrain alternatives. An incomplete list of important observations contributing to the security of the standard cosmology is

- **Big Bang Nucleosynthesis:** The cosmological formation of light atomic nuclei such as deuterium and helium-4, when the Universe was just a few seconds old.

- The accurately measured temperature and black-body form of the CMB, originating at decoupling at around 400 000 years old.

- The CMB anisotropies as we have already described.

- The clustering of galaxies as a function of epoch (including the scale of baryon acoustic oscillations).

- The clustering of dark matter as measured by gravitational lensing of light from distant galaxies. For most galaxies, the image distortions are small (known as **weak lensing**) but nevertheless nowadays readily measurable in large galaxy surveys such as the Dark Energy Survey. We discussed gravitational lensing in Chapter 10.

- The magnitude–redshift relation of distant Type Ia supernovae.

Although the collective observational support for the standard cosmological model is extremely strong, there remain some niggling discrepancies whose statistical significance is much debated. Are they due simply to over-optimistic budgeting of observational uncertainties, or do they signal the first cracks that will ultimately require the standard cosmology to be extended or even restructured? At time of writing two appear particularly noteworthy, but the situation is rapidly evolving so be sure to check for updates. One is that direct measurements of the Hubble constant, using supernovae observations calibrated using cepheid variable stars, yield an expansion rate roughly 10% higher than that inferred from the CMB, with an uncertainty small enough to put the discrepancy at the 4-sigma level. This is known as the **Hubble tension**. The second is that the amount of lithium-7 production predicted by Big Bang Nucleosynthesis is several times larger than is observed in stars. Although only about one in a billion of the cosmos's nuclei are lithium, the abundances are nevertheless well enough predicted and measured that this is a significant concern.

13.4   Accelerating cosmologies

We have already seen that the preferred cosmological model has the Universe presently accelerating. This is most simply obtained via a cosmological constant, which (at time of writing) is consistent with all reliable data. This suggests that the Universe possesses an event horizon, and that it is evolving asymptotically towards de Sitter space. However, we cannot be sure that the cosmological 'constant' does not evolve with time, when it is then referred to as dark energy; if so it might vanish again in the future or even turn negative prompting a future recollapse of the Universe.

A precedent for the former possibility lies in the theory of cosmological inflation, which postulates that the Universe underwent a transient period of accelerated expansion during its very early stages before reverting to the radiation- and matter-dominated eras. Originally proposed by Alan Guth in 1981, inflation is by now widely regarded as an essential component of the standard cosmology, because it provides a mechanism for the origin of the observed density perturbations that cause galaxy formation. These perturbations are predicted as an unavoidable consequence of quantum uncertainty during inflation, which leads to fluctuations that are caught up in the accelerated expansion and swept to the large length scales that we can presently observe astronomically. The predictions of the simplest models of inflation provide an excellent match to the very precise mapping of the primordial perturbations provided by the CMB. Moreover, inflation provides an explanation for the large-scale properties of the Universe that it has an almost smooth matter distribution and negligible spatial curvature (referred to as the horizon and flatness problems).

An alternative mechanism for cosmic acceleration, particularly in the present Universe, is that General Relativity might break down on cosmological scales. Numerous proposals exist, which are however strongly constrained by the fact that General Relativity has proven to work superbly in every regime in which it has been tested thus far.

Problems

13.1* Consider a fluid with equation of state $p = w\rho c^2$, where $-1 < w \le 1$ is a constant. Find the evolution of such a fluid in terms of the scale factor. Assuming $k = \Lambda = 0$, find the corresponding expansion rate of the Universe. For what values of w is the expansion accelerating?

13.2 Rewrite the matter- and radiation-dominated solutions for a flat Universe with $\Lambda = 0$ in terms of conformal time τ.

13.3* Consider a flat Universe containing both matter and radiation (but with $\Lambda = 0$). Show that the Friedmann equation can be written as

$$H^2 = \frac{1}{2}H_{eq}^2\left[\left(\frac{a_{eq}}{a}\right)^3 + \left(\frac{a_{eq}}{a}\right)^4\right],$$

where a_{eq} is the value of the scale factor when the densities of matter and radiation are equal and H_{eq} is the Hubble parameter at that time. By changing the time variable to conformal time τ and taking initial condition $a = 0$ at $\tau = 0$, find an expression for the scale factor evolution $a(\tau)$, such that $a = a_{eq}$ at $\tau = \tau_{eq}$. Give also a formula for τ_{eq}.

13.4 Derive the metric for an expanding de Sitter space (see Problem 12.6) using conformal time.

13.5 Find the frequency shift obtained by applying the Doppler formula for light to the recession velocity given by Hubble's Law $v = H_0\tilde{r}$. Compare it to the cosmological redshift in equation (13.20), and show that they give the same result in the limit $H_0\tilde{r}/c \to 0$, but differ in general.

13.6* By using the substitution $y = (1+z)^{-3/2}$, or otherwise, show that the age of a flat matter-dominated Universe with a cosmological constant is

$$H_0 t_0 = \frac{2}{3}\frac{1}{\sqrt{1-\Omega_m}}\sinh^{-1}\left[\sqrt{\frac{1-\Omega_m}{\Omega_m}}\right].$$

For what value of the matter density does the age equal the Hubble time?

13.7   Compare the approximate formula equation (13.33) for the particle horizon with a numerical integration of the full equation.

13.8*   Throughout this question, assume a matter-dominated Universe with $k = 0$ and $\Lambda = 0$. By considering light emitted at time t_{emit} (corresponding to a redshift z) and received at the present time t_0, show that the coordinate distance travelled by the light is given by

$$r_0 = 3ct_0 \left[1 - \frac{1}{\sqrt{1+z}} \right].$$

Derive a formula for the apparent angle subtended by an object of length l at redshift z. Find the behaviour in the limits of small and large z, and provide a physical explanation. Show that the object appears the smallest if it is located at redshift $z = 5/4$.

Chapter 14

General Relativity: The Next 100 Years

General Relativity has entered its second century · where is it now, and where will it take us?

The General Theory of Relativity is extraordinarily successful. With two parameters, the gravitational constant G and the cosmological constant Λ, it describes all known gravitational effects in the Universe with high precision, providing one accepts the existence of dark matter. It explains them in terms of a beautiful and profound picture, revealing that space and time are not just a stage on which physical processes happen, but take part in them too. It makes remarkable predictions, including black holes and gravitational waves, matches precision observations of dynamics within our Solar System and beyond it, and enables the study of the Universe as a whole. As a physical theory, its reach is simply breathtaking!

But despite its age and successes, General Relativity remains very much a living theory that continues to captivate researchers. We end with a quick look at the current frontier, which has two facets. One is to develop the theory of General Relativity, through observational tests and by theoretical progress in applying it more widely. The other is to seek to go beyond it.

14.1 Developing General Relativity

We have seen spectacular confirmation of the dynamical nature of space–time with the detection of gravitational waves from merging black holes and neutron stars. The observation using the Laser Interferometer Gravitational-wave Observatory (LIGO) in 2015 of waves emitted during the merger of a distant black-hole

Introducing General Relativity, First Edition. Mark Hindmarsh and Andrew Liddle.
© 2022 John Wiley & Sons Ltd. Published 2022 by John Wiley & Sons Ltd.
Companion website: www.wiley.com/go/hindmarsh/introducingGR

binary has galvanised scientists all over the world, both to push forward with new experiments like the space-based detector Laser Interferometer Space Antenna (LISA), and to develop new ways of testing theoretical ideas about gravity and the Universe. With gravitational waves we can potentially observe violent events in the very early Universe, perhaps even as early as the end of inflation.

On the theoretical side, one key frontier is to widen the applicability of the theory by pushing it into new regimes through numerical simulations. This field, known as numerical relativity, has been invigorated by development of techniques to safely model event horizons and prevent singularities from generating numerical artefacts. This is particularly crucial for modelling the merging black holes and neutron stars that are the source of the gravitational waves detected by LIGO, whose analysis requires accurately predicted waveforms across a wide parameter space (masses, spins, orbits). Black-hole binaries with a large mass ratio are especially challenging.

The theoretical foundations of General Relativity are also under renewed investigation. We have followed the traditional metric formulation, which assumes that the connection is symmetric, $\Gamma^{\mu}_{\nu\rho} = \Gamma^{\mu}_{\rho\nu}$, and metric compatible, $g_{\mu\nu;\rho} = 0$. This leads to the Levi-Civita connection and the association of gravity to space–time curvature. But there are alternative yet equivalent formulations which relax either or both assumptions. For example, in the so-called teleparallel formulation of General Relativity, the space–time curvature vanishes, and all gravitational effects originate from the antisymmetric part of the connection, known as **torsion**, $T^{\mu}_{\nu\rho} = \Gamma^{\mu}_{\nu\rho} - \Gamma^{\mu}_{\rho\nu}$. Other formulations relax the metric-compatibility condition. It is remarkable that these theories are indistinguishable from the usual formulation of General Relativity. Perhaps there is a hidden layer underpinning our venerable theory, and we may one day have observations to help us understand it better.

14.2   Beyond General Relativity

Will General Relativity ever be superseded, just as it supplanted Newton's theory of gravity? There are several clues which indicate that it is not the final theory of gravity.

First, there is the puzzle of the accelerating Universe. Existing observations are perfectly well satisfied through suitable choice of the cosmological constant Λ, but as we shortly see the required value sits uneasily with expectations from quantum theory. This has led many researchers to contemplate that the acceleration may instead be due to a breakdown of General Relativity, meaning that the Einstein tensor in the Einstein equation should be replaced by something else, for instance adding terms quadratic in the curvature tensors. There is no shortage of suggestions and valid options, the research topic usually referred to as **modified**

gravity, but the lack of any observational impetus for such theories has meant that so far nothing very compelling has emerged. Such theories also must tread very carefully to avoid spoiling situations where General Relativity is known to work very well (Figure 14.1).

This type of argument for seeking a replacement for General Relativity is fairly recent. But the presence of curvature singularities in solutions to Einstein's equations has troubled theorists since they were discovered. We first came across them in interpreting the Schwarzschild solution in Chapter 11, where we saw that any object falling into a black hole ends up at a point of infinite curvature after experiencing a finite proper time. What happens to the object's elementary constituents? General Relativity does not tell us. Nor does it have a satisfactory account of the very beginning of the Universe, to tell us how matter and indeed space–time itself come into being at the moment $t = 0$ in the FLRW Universe of Chapter 12.

The most powerful arguments that General Relativity cannot be complete come not from studying the theory in isolation, but from considering how it fits into the broader picture of modern physics. The known contents of the Universe are described by a quantum field theory, the Standard Model of Particle Physics. This provides a precision framework consistent with a huge range of non-gravitational experiments covering a wide range of energy scales. This suggests that General Relativity, which is also a field theory, ought to be a quantum theory too. Like the other fundamental forces, gravity would then be mediated by a particle, which is known as the **graviton**. Perhaps such a quantum theory of gravity would give a more satisfactory account of the centres of black holes and the start of the Big Bang.

If one tries to combine the Principle of General Relativity with standard techniques of quantum field theory, the first result is a spectacular success: Hawking radiation from black holes. One can picture Hawking radiation as being due to virtual particle–antiparticle pairs near the black-hole event horizon, with one of the pair disappearing into the black hole, and the other escaping to infinity. However, to make a quantum theory including gravity, one must also account for the effects of all virtual particle–antiparticle pairs, including gravitons. Methods for dealing with the effects of virtual particles are collectively known as renormalisation. Unfortunately, standard renormalisation methods do not work when applied to the simplest attempts to make a quantum theory of gravity and matter, with fundamental quantities such as the gravitational constant and particle masses becoming infinite or ill-defined.

There are a number of proposals for dealing with this failure, but no consensus. We first note that the natural scale on which the effects of any quantum theory of gravity will be felt is the Planck length ℓ_{Pl}, which is the unique quantity with the dimension of length that can be assembled from the fundamental constants G,

Planck's constant $\hbar$, and the speed of light c:

$$\ell_{Pl} \equiv \sqrt{\frac{\hbar G}{c^3}} \simeq 1.616 \times 10^{-35} \text{ m}.$$ (14.1)

We can equivalently talk about energies in terms of the Planck energy E_{Pl},

$$E_{Pl} \equiv \sqrt{\frac{\hbar c^5}{G}} \simeq 1.2 \times 10^{28} \text{ eV}.$$ (14.2)

By the standards of particle physics this is an extraordinarily high energy, such that in practice, we will never need to describe the effects of experiments probing this scale. No conceivable particle accelerator will be able to bring particles to this energy; for instance the CERN Large Hadron Collider manages to accelerate particles to 'only' 10^{13} eV. A combination of the Standard Model with General Relativity can in fact be made to work as an 'effective' quantum theory, valid only for energy scales much less than the Planck scale, by ignoring virtual particles with energies higher than the Planck energy. So do we really need a renormalisable theory of gravity?

A major problem with this viewpoint is the value of the cosmological constant. We saw in Chapter 7 that the cosmological constant could also be viewed as a form of matter with negative pressure whose magnitude is equal to the energy density. The zero-point energies of the fields of the Standard Model contribute in exactly this way, and if we add them up, we get a value 10^{122} times bigger than measured. This means that the 'bare' cosmological constant and the zero-point energies would have to cancel to an extraordinary and, to most theorists, implausible accuracy.

A fully quantum theory of gravitation has been a theoretical objective for most of General Relativity's history. However, there is no agreement on what a complete quantum theory that includes gravity might look like. Some theorists think that it is the standard renormalisation methods which are at fault, and have developing more sophisticated ones which deal with the infinities, but raise further questions. Another route is to change the variables describing the space–time from the metric tensor to the amount by which a 4-vector changes when parallel transported around a closed loop — this is known as loop quantum gravity. Perhaps the most popular is string theory, which replaces elementary particles by tiny vibrating strings and whose quantum theory (amazingly) includes General Relativity in the limit where energies are well below the Planck energy. String theory is a very rich theory, which predicts far more particles, fields, and space dimensions than the Standard Model and General Relativity combined, and the difficulty is to explain why we see only a tiny subset of the possible phenomena.

14.3 Into the future

By putting forward the Principle of General Relativity, that there is nothing special about any observer's coordinate system, Einstein was led to a remarkable outcome — the General Theory of Relativity, which reached its final form in 1915. It was developed over many years, with several wrong turnings and mistakes, and with much discussion and collaboration with scientists such as Grossmann, Hilbert, Levi-Civita, and Nordström. The theory made predictions far in advance of what could be investigated at the time: black holes and gravitational waves were already known implications by 1919, when the observation of the deflection of starlight by the Sun convinced the scientific community that the theory was correct. It took a hundred years for gravitational waves to be directly detected, and for the shadow of a black hole to be directly observed. General Relativity is so fundamental to our understanding of the Universe, and makes such extreme and counter-intuitive predictions, that researchers feel compelled to investigate it further, and to use it to look further into the Universe and so deeper into the past.

Right now, we can observe gravitational waves with frequencies in the range $100 - 300$ Hz and strain amplitudes of around 10^{-21} with ground-based gravitational wave detectors like LIGO, the Virgo interferometer, and the Kamioka Gravitational Wave Detector (KAGRA). These have already told us about extraordinary events in the distant Universe: merging black holes and neutron stars, releasing huge amounts of energy as gravitational waves.

Astronomers are using radio telescopes in Chile and the South Pole to search for indirect evidence of very long-wavelength gravitational waves in the patterns of fluctuations of the intensity and polarisation of the cosmic microwave background (CMB). Such long-wavelength gravitational waves are predicted to be produced by quantum fluctuations during inflation: their observation would be strong evidence that quantum theory must be applied to gravity as well as matter, a truly revolutionary prospect.

In the next couple of decades, the hunt for gravitational waves will intensify. Astronomers studying the CMB will combine forces in the CMB-S4 collaboration, which will increase sensitivity by around two orders of magnitude. More sensitive ground-based gravitational-wave detectors are planned, such as the Einstein Telescope, which will increase the number of merger events we can study, and promise to teach us more about the dynamics of space–time and the properties of matter in neutron stars. Detectors will also be launched into space, with sensitivity to lower frequencies and larger black holes: we already mentioned LISA, and there are also proposals TianQin and Taiji in China, and DECi-hertz Interferometer Gravitational wave Observatory (DECIGO) in Japan. Further into the future, there will be advances in technology which will extend the observable window to higher frequencies. The ultimate challenge would be to directly observe gravitational waves produced by the explosive heating of the Universe at the end of inflation, which could perhaps be at a frequency of around a gigahertz.

A century of General Relativity has brought about profound changes in our understanding of space and time, and of the Universe and its contents. Imagine what the next 100 years could bring!

Figure 14.1　Albert Einstein photographed in 1921 by Ferdinand Schmutzer. (© Österreichische Nationalbibliothek Bildarchiv.)

Advanced Topic A1

Geodesics in the Schwarzschild Space–Time

Prerequisites: Chapters 1 — 9

Now we look at geodesics in detail ·

In this chapter, we take a closer look at the geodesics in the Schwarzschild space–time, which we briefly studied in Section 9.5. We will first consider how to identify conserved quantities like energy and angular momentum. We then consider the case of massive particles, which follow geodesics that can be parametrised by proper time τ. The case of photons is similar, but requires a separate treatment as proper time can no longer be used. The result will be derivations of equations (9.28) and (9.29) for massive particles, and equations (9.32) and (9.33) for massless ones.

A1.1 Geodesics and conservation laws

Let us consider the motion of a particle of rest mass $m_0 \neq 0$. The treatment of a massless particle is similar, and the adjustments to the following argument are given later on. The 4-momentum of the particle is

$$p^\mu \equiv m_0 u^\mu, \tag{A1.1}$$

where the 4-velocity is

$$u^\mu \equiv \frac{dx^\mu}{d\tau}. \tag{A1.2}$$

Introducing General Relativity, First Edition. Mark Hindmarsh and Andrew Liddle.
© 2022 John Wiley & Sons Ltd. Published 2022 by John Wiley & Sons Ltd.
Companion website: www.wiley.com/go/hindmarsh/introducingGR

The geodesic equation (9.9) can be written in terms of the 4-momentum, using the proper time as the affine parameter, as

$$m_0 \frac{dp^\mu}{d\tau} + \Gamma^\mu_{\nu\sigma} p^\nu p^\sigma = 0. \tag{A1.3}$$

We can also write an equivalent equation for the momentum with its index lowered, which turns out to be simpler. Inspecting the covariant differential for a covariant 4-vector equation (5.19), we obtain

$$m_0 \frac{dp_\beta}{d\tau} - \Gamma^\rho_{\alpha\beta} p^\alpha p_\rho = 0. \tag{A1.4}$$

The second term can now be simplified. First, we write it as

$$\Gamma^\rho_{\alpha\beta} p^\alpha p_\rho = \frac{1}{2} \left(\partial_\alpha g_{\beta\mu} + \partial_\beta g_{\alpha\mu} - \partial_\mu g_{\alpha\beta} \right) p^\alpha p^\mu. \tag{A1.5}$$

The first and third terms in the bracket cancel when multiplied by $p^\alpha p^\mu$ and summed over α and μ, leaving us with

$$\frac{dp_\beta}{d\tau} = \frac{1}{2m_0} g_{\mu\alpha,\beta} \, p^\alpha p^\mu. \tag{A1.6}$$

The conclusion from this important equation is that if all of the metric terms are independent of a particular coordinate (x^β as written here) in some coordinate system, the right-hand side of this equation vanishes and hence the momentum component associated with that coordinate, p_β, is conserved along the particle trajectory. As a particular example, if the metric is time-independent then p_0 is conserved, representing energy conservation. If the metric is constant in some direction, it has a **Killing vector**. The proper definition of a Killing vector and its connection to conserved quantities is explored in more advanced textbooks.

A1.2 Schwarzschild geodesics for massive particles

We now seek an equation for the coordinates of a particle, $x^\mu(\tau)$, which will describe the orbit around a central body of mass M. Rather than integrating the second-order differential equation (A1.6), it is easier to use conservation laws.

As we learned in Chapter 8, the metric in the Schwarzschild space–time can be written as

$$ds^2 = -\left(1 - \frac{r_S}{r}\right) d(x^0)^2 + \left(1 - \frac{r_S}{r}\right)^{-1} dr^2 + r^2 d\theta^2 + r^2 \sin^2\theta \, d\phi^2, \tag{A1.7}$$

where we have used $x^0 = ct$ and the Schwarzschild radius $r_S \equiv 2GM/c^2$.

For the Schwarzschild space–time, the metric components are independent of both the x^0 and the ϕ coordinates, and hence the covariant 4-momentum components p_0 and p_ϕ are both conserved. To represent these two conserved quantities, we define two constants, one with units of velocity, U, and the other the specific angular momentum ℓ,

$$U = -\frac{p_0}{m_0}; \quad \ell = \frac{p_\phi}{m_0}. \tag{A1.8}$$

Additionally, we can choose coordinates such that the orbit is in the equatorial plane $\theta = \pi/2$, making θ constant and $p^\theta = 0$.

We have a further constraint to help us, which is that the 4-velocity obeys $u^\mu u_\mu = -c^2$, as we demonstrated in Section 9.1. Hence, the 4-momentum must obey

$$p^\mu p_\mu = -m_0^2 c^2. \tag{A1.9}$$

This equation involves the raised-index version of the 4-momentum, for which we have to multiply by the inverse metric, so

$$p^0 = g^{00} p_0 = m_0 \left(1 - \frac{r_S}{r}\right)^{-1} U; \tag{A1.10}$$

$$p^r = m_0 \frac{dr}{d\tau}; \tag{A1.11}$$

$$p^\phi = g^{\phi\phi} p_\phi = m_0 \frac{\ell}{r^2}. \tag{A1.12}$$

This gives

$$-m_0^2 U^2 \left(1 - \frac{r_S}{r}\right)^{-1} + m_0^2 \left(1 - \frac{r_S}{r}\right)^{-1} \left(\frac{dr}{d\tau}\right)^2 + \frac{m_0^2 \ell^2}{r^2} = -m_0^2 c^2, \tag{A1.13}$$

which simplifies to

$$\left(\frac{dr}{d\tau}\right)^2 = U^2 - \left(1 - \frac{r_S}{r}\right) \left(c^2 + \frac{\ell^2}{r^2}\right). \tag{A1.14}$$

This equation can be rearranged to give

$$\mathscr{E} = \frac{1}{2} \left(\frac{dr}{d\tau}\right)^2 + V_{\text{eff}}^{\text{GR}}(r), \tag{A1.15}$$

where $\mathscr{E} = (U^2 - c^2)/2$. The new constant $\mathscr{E}$ reduces to the Newtonian specific energy in the limit of small velocity. The general relativistic effective potential is then

$$V_{\text{eff}}^{\text{GR}}(r) = -\frac{c^2 r_{\text{S}}}{2r} + \frac{\ell^2}{2r^2} - \frac{r_{\text{S}}\ell^2}{2r^3}.$$ (A1.16)

A1.3   Schwarzschild geodesics for massless particles

A massless particle has zero rest mass and travels at the speed of light. It is not possible to define proper time for a massless particle: instead, we parametrise the particle's world line with an affine parameter λ, which has no simple physical interpretation. It is just a label along the world line of the particle. The 4-velocity is

$$u^\mu = \frac{dx^\mu}{d\lambda},$$ (A1.17)

which satisfies

$$u^\mu u_\mu = 0$$ (A1.18)

with an affine parametrisation. Any other parameter linearly related to λ is also an affine parameter.

If we have a space–time which approaches Minkowski space–time at large distances, as is the case for the Schwarzschild metric, we can define an energy E_0 as the energy of the particle at infinity, and choose the affine parameter such that the 4-momentum is

$$p^\mu = E_0 u^\mu.$$ (A1.19)

The geodesic equation for a massless particle has the same form as for a massive one, with the proper time replaced by the affine parameter. As the rest mass is zero, one examines the 4-velocity instead of the 4-momentum, which in its covariant form obeys (see equation (A1.6))

$$\frac{du_\beta}{d\lambda} = \frac{1}{2} g_{\mu\alpha,\beta} u^\alpha u^\mu.$$ (A1.20)

Just as for massive particles, there are two conserved quantities, which can be related to energy and angular momentum, resulting from the independence of the metric of the coordinates x^0 and ϕ. Hence, we can define two constants of the motion

$$U = u_0; \quad \ell = u_\phi.$$ (A1.21)

The contravariant components of the 4-velocity are

$$
\begin{aligned}
u^0 &= \frac{dx^0}{d\lambda} = g^{00} u_0 = \left(1 - \frac{r_S}{r}\right)^{-1} U\,; \\
u^r &= \frac{dr}{d\lambda}\,; \\
u^\theta &= \frac{d\theta}{d\lambda} = 0\,; \\
u^\phi &= \frac{d\phi}{d\lambda} = g^{\phi\phi} u_\phi = \frac{\ell}{r^2}\,.
\end{aligned}
\tag{A1.22}
$$

From equation (A1.18) we can write

$$
-U^2 \left(1 - \frac{r_S}{r}\right)^{-1} + \left(1 - \frac{r_S}{r}\right)^{-1} \left(\frac{dr}{d\lambda}\right)^2 + \frac{\ell^2}{r^2} = 0,
\tag{A1.23}
$$

leading to

$$
U^2 = \left(\frac{dr}{d\lambda}\right)^2 + \left(1 - \frac{r_S}{r}\right)\left(\frac{\ell^2}{r^2}\right).
\tag{A1.24}
$$

It is convenient to use the freedom to rescale the affine parameter by the constant ℓ (for compactness we will still call this rescaled parameter λ), and to define a quantity with dimensions of length $b = \ell/U$, to obtain

$$
\frac{1}{b^2} = \left(\frac{dr}{d\lambda}\right)^2 + \frac{1}{r^2}\left(1 - \frac{r_S}{r}\right).
\tag{A1.25}
$$

The quantity b has dimensions of length, and is interpreted as an impact parameter of a scattering process. Note that the equation depends only on the impact parameter, and is independent of the energy of the particle, an important difference from the massive case which means that photons of any wavelength follow the same trajectory.

Inspecting equation (A1.25), one can see that an effective potential can also be defined for a massless particle,

$$
W_{\text{eff}}(r) = \frac{1}{r^2}\left(1 - \frac{r_S}{r}\right).
\tag{A1.26}
$$

From the first of equations (A1.22) we can quickly derive an expression for the differential of the time coordinate with respect to the rescaled affine parameter,

$$\frac{dt}{d\lambda} = \frac{1}{cb}\left(1 - \frac{r_S}{r}\right)^{-1}. \tag{A1.27}$$

This equation is needed to compute the time delays of light signals in the Schwarzschild space–time, as we do in Advanced Topic A2.4.

Problems

A1.1

a) Show that the constancy of $u_\mu u^\mu$ is maintained by 4-vectors u^μ obeying the geodesic equation.

b) Let $v^\mu(x)$ be a vector field assembled from geodesics, meaning that through every space–time point x there is a geodesic $x^\mu(\lambda)$, such that

$$\frac{dv^\mu}{d\lambda} + \Gamma^\mu_{\nu\rho} v^\nu v^\rho = 0.$$

Show that

$$v^\nu v^\mu{}_{;\nu} = 0.$$

[The quantity $v^\nu v^\mu{}_{;\nu}$ is the covariant acceleration of the vector field v^μ.]

A1.2 Find the effective potentials $V_{\text{eff}}(r)$ and $W_{\text{eff}}(r)$ for orbits of massive and massless particles in the Reissner–Nordström metric, given in Problem 11.4. Write them in as compact a form as possible, using the Schwarzschild radius r_S and the Reissner–Nordström charge radius r_Q.

A1.3 Consider a Schwarzschild orbit for a massive particle with $\mathscr{E} > 0$ and $\ell/r_Sc \gg 1$. Recall the definition $U^2 = c^2 + 2\mathscr{E}$.

a) Show that if the 3-velocity at infinity is v_∞, then $v_\infty^2 = 2\mathscr{E}/(1 + 2\mathscr{E}/c^2)$ and $U/c = 1/(1 - v_\infty^2/c^2)^{1/2}$.

b) Defining an impact parameter $b = \ell/\sqrt{2\mathscr{E}}$, find the approximate impact parameter below which a particle is always drawn into the singularity, no matter its velocity at v_∞.

Advanced Topic A2

The Solar System Tests in Detail

Prerequisites: Chapters 1 — 10

Here we make predictions for the three classic Solar System tests

In this Advanced Topic, we go through the calculations behind the Solar System tests of General Relativity in more detail. The first successful prediction was the perihelion shift of Mercury, first examined in Chapter 10. When Einstein realised that his emerging theory could successfully explain this, he became certain he was on the right track.

Before studying the perihelion shift, we will review the Newtonian calculation of the motion of a body in a central potential, describing the orbit of a planet around the Sun. This calculation leads to the so-called Keplerian orbits, which are ellipses.

A2.1 Newtonian orbits in detail

We first recall from Section 9.4 the equations governing an orbit in a central potential provided by a mass M in Newtonian gravity. We have conservation of specific angular momentum

$$\ell = r^2 \frac{d\phi}{dt}, \tag{A2.1}$$

and conservation of specific energy

$$e = \frac{1}{2}\left(\frac{dr}{dt}\right)^2 + V_{\text{eff}}(r), \tag{A2.2}$$

Introducing General Relativity, First Edition. Mark Hindmarsh and Andrew Liddle.
© 2022 John Wiley & Sons Ltd. Published 2022 by John Wiley & Sons Ltd.
Companion website: www.wiley.com/go/hindmarsh/introducingGR

where V_{eff} is given by

$$V_{\text{eff}}(r) = \frac{\ell^2}{2r^2} - \frac{GM}{r}. \tag{A2.3}$$

We neglect the mass of the orbiting body in comparison to the central mass. The energy conservation equation shows that, if $e < 0$, the body moves between radii r_1 and r_2, corresponding to a bound orbit. We refer to the distance of closest approach to the central body r_1 as periapsis, and the furthest distance r_2 as apoapsis (perihelion and aphelion are reserved for bodies orbiting the Sun). As e approaches zero from below, r_2 tends to infinity; the orbit gets larger. When e becomes positive, the root r_2 goes negative, meaning that there is only one turning point in the range $r > 0$. The orbit is unbound; the orbiting body comes in from infinity and returns there.

On substituting equation (A2.3) into equation (9.24), one can derive that the angular coordinate when the body reaches a radial distance r, measured from periapsis, is

$$\phi = \frac{\ell}{\sqrt{2|e|}} \int_{r_1}^{r} \frac{dr}{r} \frac{1}{\sqrt{|(r-r_1)(r_2-r)|}}. \tag{A2.4}$$

Here we have introduced r_1 and r_2 as the roots of the equation $e = V_{\text{eff}}(r)$, which are

$$r_1 + r_2 = -\frac{GM}{e}; \tag{A2.5}$$

$$r_1 r_2 = -\frac{\ell^2}{2e}. \tag{A2.6}$$

For $e < 0$, both r_1 and r_2 are positive and give the turning points of the radial motion (see Figure A2.1). In the following, we specialise to this case.

The integral in equation (A2.4) can be evaluated by making two successive substitutions. The first one is $y = \sqrt{r_1 r_2}/r$, after which

$$\phi = \frac{\ell}{\sqrt{2|e|}} \frac{1}{\sqrt{r_1 r_2}} \int_{y_2}^{y} dy \frac{1}{\sqrt{(y_1-y)(y-y_2)}}, \tag{A2.7}$$

where $y_1 = \sqrt{r_2/r_1}$ and $y_2 = \sqrt{r_1/r_2}$. We can then define

$$y_{\pm} = \frac{1}{2}(y_1 \pm y_2), \tag{A2.8}$$

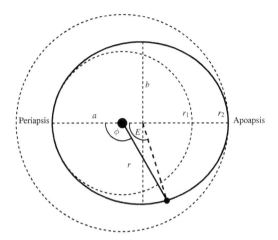

Figure A2.1 The Keplerian elliptical orbit predicted by Newtonian gravity is shown as the solid line, with a the semi-major axis and b the semi-minor axis. The central mass is the large solid dot, and the dashed circles indicate the distances of closest approach (periapsis) and furthest excursion (apoapsis). The angle ϕ indicates the angular position in the orbit relative to periapsis, with the orbiting body at distance r from the central body. The angle E with respect to the centre of the ellipse is known as the eccentric anomaly.

make the substitution $x = (y - y_+)/y_-$, and use equation (A2.6) to obtain

$$\phi = \int_{-1}^{x} dx \frac{1}{\sqrt{1 - x^2}} = \cos^{-1}(x) + \pi. \tag{A2.9}$$

Apoapsis is at $x = 1$, and so the angular displacement between periapsis and the next apoapsis $\Delta\phi$ is precisely π. The next periapsis is back at $x = -1$, where the angular displacement is precisely 2π. Keplerian orbits therefore close.

We can invert equation (A2.9) to give the orbital distance r as a function of angle ϕ, firstly by recovering

$$y = y_+ - y_- \cos \phi \tag{A2.10}$$

and then rewriting in terms of new parameters

$$a = \frac{r_1 + r_2}{2}; \quad \varepsilon = \frac{r_2 - r_1}{r_2 + r_1} \tag{A2.11}$$

to obtain

$$r = \frac{a(1-\varepsilon^2)}{1+\varepsilon\cos\phi},$$ (A2.12)

which is equation (9.27). This is the equation of an ellipse[1] with semi-major axis a and eccentricity ε. The semi-minor axis of the ellipse is $b = \sqrt{r_1 r_2}$.

To find how the body moves around its orbit, we integrate the equation $\ell = r^2 d\phi/dt$, changing integration variable, to find

$$\ell(t-t_1) = \int_{r_1}^{r} r^2 \frac{d\phi}{dr} dr = \frac{\ell}{\sqrt{2|e|}} \int_{r_1}^{r} dr' \frac{r'}{\sqrt{(r'-r_1)(r_2-r')}},$$ (A2.13)

where t_1 is the time at periapsis. This integral can be done with the substitution

$$r = a(1-\varepsilon\cos E),$$ (A2.14)

obtaining

$$t - t_1 = \frac{a}{\sqrt{2|e|}}(E - \varepsilon\sin E).$$ (A2.15)

The period is the time taken to return to the same radius, which means the time taken for E to increase by 2π, or

$$T = 2\pi\frac{a}{\sqrt{2|e|}}.$$ (A2.16)

Eliminating e using equation (A2.5) leads to

$$T^2 = \frac{4\pi^2 a^3}{2GM}.$$ (A2.17)

This is Kepler's Third Law; for a fixed central mass M, the orbital period squared is proportional to the cube of semi-major axis of the orbit.

Using the period, we can rewrite the time equation as

$$t = \frac{T}{2\pi}(E - \varepsilon\sin E).$$ (A2.18)

It is not possible to invert this equation to give a simple form for $r(t)$. The variable

---

[1] It is also the equation of a hyperbola if $\varepsilon > 1$ and $a < 0$, which describes the motion of unbound orbits. See Problem A2.1 for how this is done.

E is grandly named the eccentric anomaly, which is the angle the body subtends at the centre of the ellipse. The angle ϕ is also known as the true anomaly. Some more algebra shows they are related by

$$\sin E = \frac{\sin \phi \sqrt{1 - \varepsilon^2}}{1 + \varepsilon \cos \phi} . \tag{A2.19}$$

These formulae also apply to two bodies with mass m_1 and m_2 orbiting each other, without the assumption that one mass is much greater than the other. In this case, $M = m_1 + m_2$, $\mathbf{r}$ is the displacement vector between the two bodies, and each body orbits the centre of mass at displacements $\mathbf{r}_2 = \mathbf{r} m_1 / M$ and $\mathbf{r}_1 = -\mathbf{r} m_2 / M$. The two-body problem is not exactly solvable in General Relativity.

A2.2 Perihelion shift in General Relativity

Our aim is in this section is to find the change in the angular coordinate between perihelion and aphelion $\Delta \phi$, for the orbits predicted by General Relativity. The perihelion shift, or precession, is defined to be twice this change minus 2π, or

$$\Delta \phi = \pi + \frac{\delta \phi_{\text{prec}}}{2} . \tag{A2.20}$$

We saw in Section A2.1 that in Keplerian orbits, the angular distance between perihelion and the next aphelion is precisely π, meaning $\delta \phi_{\text{prec}} = 0$.

In the General Relativistic case, in place of equation (A2.4) we need instead to integrate the differential equation

$$\frac{d\phi}{dr} = \frac{\ell}{\sqrt{2|\mathcal{E}|} r} \frac{1}{\sqrt{(r - r_0')(r - r_1')(r_2' - r)}} , \tag{A2.21}$$

where r_0', r_1', and r_2' are the roots of the equation $\mathcal{E} = V_{\text{eff}}^{\text{GR}}(r)$, which can be turned into the equation for the roots of a cubic polynomial

$$r^3 - \frac{2GM}{|\mathcal{E}|} r^2 + \frac{\ell^2}{2|\mathcal{E}|} r - \frac{GM\ell^2}{|\mathcal{E}|c^2} = 0 . \tag{A2.22}$$

This equation can be rewritten as

$$r(r - r_1)(r_2 - r) + \frac{1}{2} r_S r_1 r_2 = 0 , \tag{A2.23}$$

where r_S is again the Schwarzschild radius, and r_1 and r_2 are the Newtonian turning points given in equations (A2.5) and (A2.6), with the Newtonian specific energy e replaced by the analogous relativistic quantity

$$r_1 + r_2 \;\; = \;\; \frac{c^2 r_S}{2|\mathcal{E}|}; \tag{A2.24}$$

$$r_1 r_2 \;\; = \;\; \frac{\ell^2}{2|\mathcal{E}|}. \tag{A2.25}$$

Hence, if the Newtonian turning points are well outside the Schwarzschild radius, and writing $\delta r_{1,2} = r'_{1,2} - r_{1,2}$, we find

$$r'_0 \simeq r_S; \quad \delta r_1 \simeq -r_S \frac{r_2}{(r_2 - r_1)}; \quad \delta r_2 \simeq +r_S \frac{r_1}{(r_2 - r_1)}. \tag{A2.26}$$

Using the fact that $r'_0 \ll r$ for orbits in the Solar System, we can expand the differential equation for the angular coordinate as

$$\frac{d\phi}{dr} = \frac{\ell}{\sqrt{2|\mathcal{E}|}} \frac{1}{r\sqrt{(r - r'_1)(r'_2 - r)}} \left(1 + \frac{1}{2} \frac{r'_0}{r} \right). \tag{A2.27}$$

Hence, we can write

$$\Delta\phi = \sqrt{r_1 r_2} \left(I_A + \frac{1}{2} I_B \right), \tag{A2.28}$$

where

$$I_A = \int_{r'_1}^{r'_2} \frac{dr}{r} \frac{1}{\sqrt{(r - r'_1)(r'_2 - r)}}; \quad I_B = \int_{r'_1}^{r'_2} \frac{dr}{r^2} \frac{r'_0}{\sqrt{(r - r'_1)(r'_2 - r)}}. \tag{A2.29}$$

We see that I_A is the same integral as the Newtonian case, equal to $\pi/\sqrt{r'_1 r'_2}$. Hence,

$$\Delta\phi = \pi + \frac{1}{2} \delta\phi^A_{\text{prec}} + \frac{1}{2} \delta\phi^B_{\text{prec}}, \tag{A2.30}$$

with

$$\delta\phi^A_{\text{prec}} = 2\pi \left(\frac{\sqrt{r_1 r_2}}{\sqrt{r'_1 r'_2}} - 1 \right). \tag{A2.31}$$

The integral for $\delta\phi^B_{\text{prec}}$ can be evaluated with the same substitutions as in the

Newtonian case, to give

$$I_B = \frac{\pi}{2} \frac{r_0'(r_1' + r_2')}{(r_1' r_2')^{\frac{3}{2}}}.$$ (A2.32)

Hence, to first order in r_S,

$$\delta\phi_{\text{prec}}^B \simeq \pi \frac{r_S(r_1 + r_2)}{(r_1 r_2)}.$$ (A2.33)

Using the expressions for r_0', r_1', and r_2', and again working to first order in r_S,

$$\begin{aligned} \delta\phi_{\text{prec}}^A &= 2\pi \left(-\frac{1}{2} \frac{\delta r_1}{r_1} - \frac{1}{2} \frac{\delta r_2}{r_2} \right); \\ &= \frac{\pi}{2} \frac{r_S(r_1 + r_2)}{r_1 r_2}. \end{aligned}$$ (A2.34)

Hence, the total perihelion shift in one orbit of a test body in the Schwarzschild metric is

$$\delta\phi_{\text{prec}} = \delta\phi_{\text{prec}}^A + \delta\phi_{\text{prec}}^B = \frac{3\pi}{2} \frac{r_S(r_1 + r_2)}{r_1 r_2}.$$ (A2.35)

By defining an average radius $\bar{r} = 2r_1 r_2 / (r_1 + r_2)$, we arrive at

$$\delta\phi_{\text{prec}} = 3\pi \frac{r_S}{\bar{r}}.$$ (A2.36)

One can rewrite this equation in terms of more astronomically relevant orbital parameters, namely the semi-major axis $a = (r_1 + r_2)/2$, the eccentricity $\varepsilon = |r_1 - r_2|/(r_1 + r_2)$, and the period T given in equation (A2.17). The relativistic contribution to the perihelion shift is then

$$\delta\phi_{\text{prec}} = 6\pi \left(\frac{2\pi}{cT} \right)^2 \frac{a^2}{1 - \varepsilon^2}.$$ (A2.37)

Alternatively, one can eliminate the semi-major axis in favour of the period and the total mass, giving

$$\delta\phi_{\text{prec}} = 6\pi \left(\frac{\pi r_S}{cT} \right)^{2/3} \frac{1}{1 - \varepsilon^2}.$$ (A2.38)

A2.3   Light deflection

In this section, we calculate the deflection angle of light ray as it passes by a gravitating body. As the photon comes in from infinity, passes by the object, and retreats to infinity, the angular coordinate changes by an amount $\Delta\phi$ which we will calculate from the geodesic equation. The deflection angle is the difference between $\Delta\phi$ and π, the angle subtended at the origin by an infinite straight line.

$$\delta\phi_{\text{def}} = \Delta\phi - \pi .   \tag{A2.39}$$

This angle can be computed by taking the ratio of the equations for $dr/d\lambda$ and $d\phi/d\lambda$, derived in Advanced Topic A1:

$$\frac{d\phi}{d\lambda} = \frac{1}{r^2} ; \qquad \frac{dr}{d\lambda} = \left[\frac{1}{b^2} - \frac{1}{r^2}\left(1 - \frac{r_S}{r}\right) \right]^{1/2} ,   \tag{A2.40}$$

where r_S is the Schwarzschild radius of the massive body around which the light is travelling, and b is the impact parameter. The choice of the positive root just decides which way round the body the photon travels.

We can calculate the angular coordinate change by integrating

$$\frac{d\phi}{dr} = \frac{d\phi/d\lambda}{dr/d\lambda} = \frac{1}{r^2}\left[\frac{1}{b^2} - \frac{1}{r^2}\left(1 - \frac{r_S}{r}\right) \right]^{-1/2} .   \tag{A2.41}$$

The limits of the integration are $r = \infty$ and the distance of closest approach $r = r_1$, which is a solution to

$$\frac{1}{b^2} - \frac{1}{r_1^2}\left(1 - \frac{r_S}{r_1}\right) = 0 .   \tag{A2.42}$$

Defining $y = r_S/r$, we have

$$\Delta\phi = \int_0^{y_1} dy \frac{1}{\sqrt{y_1^2 - y^2 - (y_1^3 - y^3)}} ,   \tag{A2.43}$$

where $y_1 = r_S/r_1$. This integral can be expressed analytically in terms of an elliptic function, but if the distance of closest approach is much bigger than the Schwarzschild radius (as it will be for all astronomical bodies except black holes and neutron stars), we can also perform an expansion in powers of y_1. With the change of variable $y = y_1 \cos\alpha$, and expanding the denominator to first order in

y_1, we find

$$
\Delta\phi \simeq 2\int_0^{\pi/2} d\alpha \left[1 + \frac{y_1}{2}\left(\cos\alpha + \sec^2(\alpha/2)\right)\right],
$$
$$
= \pi + 2y_1.
\tag{A2.44}
$$

If the impact parameter is large compared with the Schwarzschild radius, $b \gg r_S$, an approximate solution to equation (A2.42) is $r_1 \simeq b$. In this case, the General Relativistic prediction for the deflection angle of the photon is

$$
\delta\phi_{\text{def}} \simeq 2\frac{r_S}{b}.
\tag{A2.45}
$$

Actually, if one treats the photon as a ballistic particle of velocity c, a deflection can be obtained in Newtonian gravity, which is precisely half the General Relativity value. The extra contribution in GR can be traced to the light responding to the spatial curvature as well as the analogue of the Newtonian potential.

A2.4 Time delay

In order to calculate the light travel time from Earth, we will need $dr/d\lambda$ and $dt/d\lambda$ from our study of photon orbits in Advanced Topic A1,

$$
\frac{dt}{d\lambda} = \frac{1}{cb}\left(1 - \frac{r_S}{r}\right)^{-1}; \qquad \frac{dr}{d\lambda} = \left[\frac{1}{b^2} - \frac{1}{r^2}\left(1 - \frac{r_S}{r}\right)\right]^{1/2},
\tag{A2.46}
$$

where b is the impact parameter of the geodesic. Taking the ratio of the two equations, we find

$$
\frac{dt}{dr} = \frac{1}{cb}\left(1 - \frac{r_S}{r}\right)^{-1}\left[\frac{1}{b^2} - \frac{1}{r^2}\left(1 - \frac{r_S}{r}\right)\right]^{-1/2}.
\tag{A2.47}
$$

The light travel time from the Earth to the point closest to the Sun is the integral between r_E and r_1. The distance of closest approach r_1 was already found in Section A2.3.

Defining a dimensionless variable $x = r/r_1$, we have

$$
\Delta t_E = \frac{r_1^2}{cb}\int_1^{x_E} dx \left(1 - \frac{x_S}{x}\right)^{-1}\left[1 - \frac{1}{x^2} - \frac{x_S}{2}\left(1 - \frac{1}{x^3}\right)\right]^{-1/2},
\tag{A2.48}
$$

where $x_S = r_S/r_1$ and $x_E = r_E/r_1$.

In the Solar System $x_S \ll 1$, as even light grazing the Sun ($R_\odot \simeq 7 \times 10^5$km) is much further than a Schwarzschild radius away (for the Sun, $r_S \simeq 3$km). The radius of the Earth's orbit ($r_E \sim 10^8$km) is much larger than the Solar radius, so we may take $x_E \gg 1$. The radii of the orbits of other planets, even Mercury, are also large compared with the radius of the Sun.

With the smallness of x_S, it is a good approximation to make an expansion in powers of x_S, leading to

$$\Delta t_E = \frac{r_1^2}{cb} \int_1^{x_E} dx \frac{x}{\sqrt{x^2-1}} \left(1 + \frac{x_S}{x} + \frac{x_S}{2}\frac{1-x^{-3}}{1-x^{-2}} \right). \tag{A2.49}$$

The integration may be performed with the substitution $x = \cosh\eta$, leading to

$$\Delta t_E = \frac{r_1^2}{cb} \left[\left(1+\frac{x_S}{2}\right)\sinh\eta_E + x_S\eta_E + \frac{x_S}{2}\tanh\frac{\eta_E}{2} \right]. \tag{A2.50}$$

Using the fact that $x_E \gg 1$ and $r_1 \simeq b$, one can show that

$$\Delta t_E \simeq \frac{1}{c}\left(r_E + r_S \ln\frac{2r_E}{r_1} \right). \tag{A2.51}$$

The first term in the brackets is just the travel time in Minkowski space–time, in the approximation that $r_E \gg r_1$. The General Relativistic time delay is the second term.

The total travel time for a signal to travel from the Earth to the planet and back is $\Delta t = 2(\Delta t_E + \Delta t_P)$, where Δt_P is the light travel time between the distance of closest approach and the planet at radius r_P. The accumulated time delay is therefore

$$\delta t_{\text{del}} = \frac{2r_S}{c}\ln\frac{4r_E r_P}{r_1^2}. \tag{A2.52}$$

Problems

A2.1   In Section A2.1, we studied the case of bound Keplerian orbits, showing that the distance from the central body r and the angle from periapsis were related by

$$r = \frac{a(1-\varepsilon^2)}{1+\varepsilon\cos\phi},$$

with

$$a = \frac{r_1 + r_2}{2}, \quad \varepsilon = \frac{r_2 - r_1}{r_2 + r_1},$$

where r_1 and r_2 are the roots of the equation $e = V_{\mathrm{eff}}(r)$. Show that the same formula applies for unbound orbits, where $e > 0$. It is convenient to choose the roots so that $r_2 < 0$, with $|r_2| > r_1$, so that r_1 is the distance at periapsis. This is the equation for a hyperbola, for which $a < 0$ and $\varepsilon > 1$.

A2.2   Calculate the light-deflection angle in Newtonian gravity, treating the photons as ordinary particles travelling at the speed of light, and assuming that the deflection angle is small. You should find that you get one-half the light-deflection angle according to General Relativity.

[Hint: you can do this two ways, either by calculating the impulse given to the particle in the perpendicular direction to its initial velocity, or by calculating the parameters of the hyperbolic orbit.]

A2.3   The binary pulsar PSR 1913+16 consists of two neutron stars orbiting with period $T = 0.323$ days, eccentricity $\varepsilon = 0.617$, and rate of periastron advance $\dot{\omega} = 4.23 \deg \mathrm{yr}^{-1}$. Estimate the total mass of the system.

Advanced Topic A3

Weak Gravitational Fields and Gravitational Waves

Prerequisites: Chapters 1 — 10

*We look closely at space–times that are almost flat · this leads to the
prediction of gravitational waves · gravitational waves are produced
only when an object's quadrupole moment varies*

A3.1 Nearly-flat space–times

Weak gravitational fields are those in which the space–time is nearly flat. This
means that coordinates exist such that we can write

$$g_{\mu\nu} = \eta_{\mu\nu} + h_{\mu\nu}, \qquad (A3.1)$$

where $\eta_{\mu\nu}$ is the metric of Special Relativity, $\text{diag}(-1, 1, 1, 1)$, and

$$|h_{\mu\nu}| \ll 1. \qquad (A3.2)$$

[NB: the inverse statement, that the metric is not nearly flat if the metric components are not close to those of $\eta_{\mu\nu}$, is *not* true, as we can use any coordinate
system we like to describe a flat or nearly-flat space–time. There only needs to
exist a particular coordinate system with this property.] In the weak-gravity limit,
we argue that non-linear terms, of order 'h^2' and above, can be dropped.

A useful interpretation of this formula is to pretend that actually space–time is
flat, with the $\eta_{\mu\nu}$ metric, and $h_{\mu\nu}$ is a rank-two tensor defined in that flat space–
time. Indeed, it turns out that $h_{\mu\nu}$ transforms according to the Lorentz transfor-

Introducing General Relativity, First Edition. Mark Hindmarsh and Andrew Liddle.
© 2022 John Wiley & Sons Ltd. Published 2022 by John Wiley & Sons Ltd.
Companion website: www.wiley.com/go/hindmarsh/introducingGR

mation. The indices of $h_{\mu\nu}$ can be raised and lowered using $\eta_{\mu\nu}$, rather than the full metric, as the correction terms would be higher order.

In the linear approximation, one can show that (see Problem A3.2)

$$R_{\mu\nu\alpha\beta} = \frac{1}{2}\left(h_{\mu\beta,\nu,\alpha} + h_{\nu\alpha,\mu,\beta} - h_{\mu\alpha,\nu,\beta} - h_{\nu\beta,\mu,\alpha}\right),\qquad(A3.3)$$

where all the derivatives can be taken to be partial rather than covariant as the covariant terms will all be higher than linear order. None of the terms with products of Christoffel symbols contribute as they are automatically at least second-order in $h^{\mu\nu}$.

We now wish to use this Ansatz for the metric in the Einstein equations, $G_{\mu\nu} = \kappa T_{\mu\nu}$. The computation is subtle as one has to allow for coordinate transformations that alter $h_{\mu\nu}$ while leaving it small. To be precise, coordinate transformations of the form $x^{\mu} \to x^{\mu} + \xi^{\mu}$, with ξ small, change h according to the rule

$$h_{\mu\nu} \to h_{\mu\nu} + \partial_{\mu}\xi_{\nu} + \partial_{\nu}\xi_{\mu},\qquad(A3.4)$$

which is a different nearly-flat metric.

It is particularly convenient to choose coordinates for which

$$h^{\nu}_{\mu,\nu} - \frac{1}{2}h^{\alpha}_{\alpha,\nu} = 0,\qquad(A3.5)$$

a choice known as the Lorenz gauge.[1] It is also useful to define

$$\bar{h}^{\mu\nu} = h^{\mu\nu} - \frac{1}{2}\eta^{\mu\nu}h^{\alpha}_{\alpha},\qquad(A3.6)$$

which is called the 'trace reverse' of $h^{\mu\nu}$, on the grounds that its trace is the negative of the trace of $h^{\mu\nu}$.

The weak-field Einstein equations can then be shown to be (see Problem A3.2)

$$\Box\bar{h}^{\mu\nu} = -2\kappa T^{\mu\nu}.\qquad(A3.7)$$

---

[1] Whose poor inventor Ludwig Lorenz is forever being confused with Hendrik Lorentz (of Lorentz transformation fame) or Edward Lorenz (of Lorenz attractor fame). Lorenz worked with electromagnetism, and one should really call this the Harmonic or de Donder gauge $\Gamma^{\mu}_{\nu\rho}g^{\nu\rho} = 0$, applied to the linearised theory. In this linear approach to the Einstein equation, the word gauge is used interchangeably with coordinate choice, by analogy with electromagnetism.

One can also show (see Problem A3.3) that the original metric perturbation obeys

$$\Box h^{\mu\nu} = -2\kappa \left(T^{\mu\nu} - \frac{1}{2}\eta^{\mu\nu} T \right), \tag{A3.8}$$

where $T \equiv T^{\mu}{}_{\mu}$ is the trace of the energy–momentum tensor. We saw in Chapter 7 that $\kappa = 8\pi G/c^4$, by comparison to the Newtonian limit.

In fact, the gauge condition (A3.5) does not fully fix the coordinate system, as one can quickly verify that

$$\partial_\mu \bar{h}^{\mu\nu} \to \partial_\mu \bar{h}^{\mu\nu} + \Box \xi^\nu. \tag{A3.9}$$

Hence,, we still have the freedom to make coordinate transformations which satisfy $\Box \xi^\nu = 0$. This remaining freedom, known as a harmonic coordinate transformation, will be useful in the next section.

A3.2 Gravitational waves

We now examine situations where the gravitational field is weak, but not stationary, and in vacuum, i.e. the energy–momentum tensor vanishes. Hence, the equation for the metric perturbation is

$$\Box \bar{h}^{\mu\nu} \equiv \left(-\frac{1}{c^2}\frac{\partial^2}{\partial t^2} + \nabla^2 \right) \bar{h}^{\mu\nu} = 0, \tag{A3.10}$$

which can be recognised as a wave equation.

Its solutions are of the form

$$\bar{h}^{\mu\nu}(x) = A^{\mu\nu} \exp(ik_\rho x^\rho) + A^{*\mu\nu} \exp(-ik_\rho x^\rho), \tag{A3.11}$$

where k_ρ is the (constant) 4-dimensional wave vector and $A^{\mu\nu}$ is a tensor with constant components. This is a plane gravitational wave, whose angular frequency is given by $\omega = k^0 c$ and spatial wave vector by k^i. The wave equation demands that $\omega^2 = k^2 c^2$, where $k = |\mathbf{k}|$.

The tensor $A^{\mu\nu}$ cannot be chosen freely, as the field $\bar{h}^{\mu\nu}$ must satisfy the Lorenz gauge condition equation (A3.5). This is assured if

$$A^\nu{}_\mu k_\nu = 0. \tag{A3.12}$$

Without loss of generality, we can suppose that the gravitational wave is propagating in the z direction, with wave vector $k^\mu = (k, 0, 0, k)$. It is not too hard to show that the choice

$$A_{\mu\nu} = \begin{pmatrix} 0 & 0 & 0 & 0 \\ 0 & A_{xx} & A_{xy} & 0 \\ 0 & A_{xy} & -A_{xx} & 0 \\ 0 & 0 & 0 & 0 \end{pmatrix}, \tag{A3.13}$$

satisfies the Lorenz gauge condition. It is more difficult to show that, using the remaining freedom to make harmonic coordinate transformations, all choices of the tensor A satisfying the Lorenz gauge condition can be transformed into equation (A3.13). Once this is done, we have a coordinate system with the catchy name 'transverse–traceless gauge'.

Thus we see that the solution in fact has only two ways to oscillate, known as the polarisations of the gravitational wave. It is conventional to refer to them with the symbols $+$ and $\times$, so that the oscillation amplitudes $A_{xx} \equiv A_+$ and $A_{xy} \equiv A_\times$. We can write the general solution in terms of polarisation basis tensors, which for the chosen wave vector $k^\mu = (k, 0, 0, k)$ are

$$\varepsilon_+^{\mu\nu} = \begin{pmatrix} 0 & 0 & 0 & 0 \\ 0 & 1 & 0 & 0 \\ 0 & 0 & -1 & 0 \\ 0 & 0 & 0 & 0 \end{pmatrix}, \quad \varepsilon_\times^{\mu\nu} = \begin{pmatrix} 0 & 0 & 0 & 0 \\ 0 & 0 & 1 & 0 \\ 0 & 1 & 0 & 0 \\ 0 & 0 & 0 & 0 \end{pmatrix}. \tag{A3.14}$$

These tensors can be written in terms of orthogonal basis vectors $\mathbf{x}$, $\mathbf{y}$, and $\mathbf{z}$ as

$$\varepsilon_+^{ij}(\hat{\mathbf{z}}) = \hat{x}^i \hat{x}^j - \hat{y}^i \hat{y}^j; \quad \varepsilon_\times^{ij}(\hat{\mathbf{z}}) = \hat{x}^i \hat{y}^j + \hat{y}^i \hat{x}^j. \tag{A3.15}$$

Note that the trace reverses of the basis tensors satisfy $\bar{\varepsilon}_A^{\mu\nu} = \varepsilon_A^{\mu\nu}$ (with the index A taking the values $+$ and $\times$), as the basis tensors are already traceless. Hence, the plane-wave solution can be expressed as

$$\bar{h}^{\mu\nu}(x) = \left(h_+ \varepsilon_+^{\mu\nu} + h_\times \varepsilon_\times^{\mu\nu} \right) e^{ik \cdot x} + \text{c.c.}. \tag{A3.16}$$

By Fourier's theorem, the general solution is a superposition of these plane-wave solutions with different wavenumbers, which can be written

$$\bar{h}^{\mu\nu}(x) = \sum_{A=+,\times} \int \frac{d^3 k}{(2\pi)^3} \left(h_A(\mathbf{k}) \varepsilon_A^{\mu\nu}(\mathbf{k}) e^{ik \cdot x} + h_A^*(\mathbf{k}) \varepsilon_A^{\mu\nu}(\mathbf{k}) e^{-ik \cdot x} \right). \tag{A3.17}$$

An effect of a gravitational wave is to change distances transverse to the propagation direction of the wave, periodically stretching them in one direction and squeezing them in another. Figure A3.1 sketches the effect of one cycle of a grav-

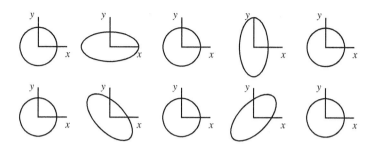

Figure A3.1 The effect of a gravitational wave passing through a ring of free test masses, greatly exaggerated. Top is one cycle of a wave with + polarisation, travelling in the z direction, perpendicular to the plane of the masses. Bottom shows the effect of a wave with × polarisation.

itational wave of either polarisation passing through a ring of free test masses (i.e. travelling into or out of the page). The different polarisations perform their stretching and squeezing along directions which are oriented at an angle of $\pi/4$ to each other. For example, a wave with + polarisation travelling in the z direction passing though a ring of test masses of diameter L in the x–y plane, causes the spatial separation of the masses lying on the x-axis to change from L to $L + \Delta L$, where

$$L + \Delta L(t) = \int_{-L/2}^{L/2} dx \sqrt{(\delta_{ij} + h_{ij}(x))\hat{x}^i \hat{x}^j}. \tag{A3.18}$$

If the wavelength of the gravitational wave is much greater than L, the metric perturbation does not change significantly with x between the test masses, and we find

$$\frac{\Delta L}{L} \simeq \frac{1}{2} h_{ij}(t, \mathbf{x}_0)\hat{x}^i \hat{x}^j, \tag{A3.19}$$

where $\mathbf{x}_0$ is the centre of the ring, which we can take to be the origin of the coordinates.

The relative change of length can be viewed as a shear strain, just as in a material: this is a strain in space itself.[2] We see that for a wave with both polarisations present, the strain in the x direction is

$$\frac{\Delta L}{L} \simeq \frac{1}{2} h_+ e^{i\omega t} + \text{c.c.}. \tag{A3.20}$$

---

[2]The idea of space having a strain, and that this strain could propagate from one place to another transporting energy and momentum, took some time to establish. Indeed, Einstein himself doubted the physical nature of gravitational waves for a while.

That is, the test masses on the x-axis do not respond to a wave with polarisation axes at an angle $\pi/4$ to the Cartesian axes.

One can similarly establish that test particles at $(x, y) = (\pm 1, \pm 1)/\sqrt{2}$ do not respond to gravitational waves travelling in the z direction with $+$ polarisation.

A3.3   Sources of gravitational waves

The source of gravitational waves is the transverse–traceless (TT) part of the energy–momentum tensor. In TT coordinates, this is contained in the spatial parts of the energy–momentum tensor, T_{ij} with $i \neq j$; in other words, the shear stress.

The solutions can be studied without worrying about selecting precisely the right components at this stage. We will write

$$\Box \bar{h}^{ij} = -\frac{16\pi G}{c^4} T^{ij}, \qquad (A3.21)$$

and use our experience with solutions of the electromagnetic wave equation to write

$$\bar{h}^{ij}(t, \mathbf{x}) = -\frac{4G}{c^4} \int d^3x' \frac{T^{ij}(t_{\mathrm{r}}, \mathbf{x}')}{|\mathbf{x} - \mathbf{x}'|}, \qquad (A3.22)$$

where the source is considered at the retarded time it $t_{\mathrm{r}} = t - |\mathbf{x} - \mathbf{x}'|/c$.

We suppose that the radiation is emitted by a system of finite extent, and that we are interested in the metric perturbation at a distance much greater than the extent of the system. In this case, $r = |\mathbf{x} - \mathbf{x}'|$ is approximately constant as one integrates over the source position $\mathbf{x}'$, and we can write

$$\bar{h}^{ij}(t, \mathbf{x}) \simeq -\frac{4G}{rc^4} \int d^3x' T^{ij}(t_{\mathrm{r}}, \mathbf{x}'), \qquad (A3.23)$$

with $t_{\mathrm{r}} = t - r/c$.

The energy–momentum conservation equations in the weak field limit are just the Minkowski space conservation equations $T^{\mu\nu}{}_{,\nu}$, which we write as

$$\frac{1}{c}\frac{\partial}{\partial t} T^{00} + \frac{\partial}{\partial x^i} T^{i0} = 0; \qquad (A3.24)$$

$$\frac{1}{c}\frac{\partial}{\partial t} T^{0j} + \frac{\partial}{\partial x^i} T^{ij} = 0. \qquad (A3.25)$$

Differentiating the first equation with respect to time, and using the second equation, we find

$$\frac{1}{c^2}\frac{\partial^2}{\partial t^2}T^{00} = \frac{\partial^2}{\partial x^i \partial x^j}T^{ij}.$$

(A3.26)

Multiplying both sides by $x^k x^l$ and integrating over all space, we have

$$\frac{1}{c^2}\frac{\partial^2}{\partial t^2}\int d^3x\, x^k x^l T^{00} = \int d^3x\, x^k x^l \frac{\partial^2}{\partial x^i \partial x^j}T^{ij}.$$

(A3.27)

Integrating the right-hand side by parts, and using the assumption that the energy–momentum tensor is non-zero in a region limited in spatial extent, we can drop the boundary terms to find

$$\frac{d^2}{dt^2}I^{kl} = 2\int d^3x\, T^{kl},$$

(A3.28)

where I^{kl} is the second moment of the mass density,

$$I^{kl} = \frac{1}{c^2}\int d^3x\, x^k x^l T^{00}.$$

(A3.29)

Thus, we arrive at

$$\bar{h}^{ij}(r,\mathbf{x}) = -\frac{2G}{rc^4}\ddot{I}^{ij}(t_r).$$

(A3.30)

This expression is not yet TT, as the second moment is not traceless, nor in general transverse to the vector along direction of propagation $\mathbf{n}$. We can take the source position relative to its centre of mass $\mathbf{x}_{cm}$, in which case $\mathbf{n}$ is defined through $\mathbf{x} - \mathbf{x}_{cm} = r\mathbf{n}$.

A detailed calculation shows that a coordinate transformation can always be made which brings $\bar{h}^{ij}$ into TT form. We first introduce some notation for the spatially trace-free form of the second moment of the mass distribution,

$$\not{I}^{ij} = I^{ij} - \frac{1}{3}I^k_k \delta^{ij}.$$

(A3.31)

This is the **quadrupole moment** of the mass distribution.

We then define a tensor which projects out the component of a vector parallel to $\mathbf{n}$, thus leaving it transverse:

$$P^{ij} = \delta^{ij} - n^i n^j.$$

(A3.32)

We need something which does the same for two-index tensors A_{ij}, that is, produces a tensor A^{ij}_{TT} which is transverse (so that $A^{ij}_{TT} n_j = 0$) and traceless ($A^i_{TT\,i} = 0$). Such an object must have four indices, and it is straightforward but lengthy to check that the following object will have this effect when contracted with a two-index tensor on any pair of indices:

$$\Lambda^{ij}{}_{kl}(\mathbf{n}) = \frac{1}{2}\left(P^i_k P^j_l + P^i_l P^j_k - P^{ij} P_{kl} \right). \tag{A3.33}$$

Hence, the TT quadrupole tensor is

$$I^{ij}_{TT} = \Lambda^{ij}{}_{kl}(\mathbf{n}) I^{ij}, \tag{A3.34}$$

giving the TT form of the metric perturbation at a large distance r in a direction $\mathbf{n}$ away from a gravitational wave source as

$$\bar{h}^{ij}_{TT}(r,\mathbf{n}) = -\frac{2G}{rc^4} \ddot{I}^{ij}_{TT}(t_r). \tag{A3.35}$$

This equation is known as the **quadrupole formula**.

Problems

A3.1   Consider the transverse–traceless projector (A3.33). Show that it is
a) transverse, i.e. that $n_i \Lambda^{ij}{}_{kl}(\mathbf{n}) = 0$ and $n_j \Lambda^{ij}{}_{kl}(\mathbf{n}) = 0$,
b) traceless, i.e. that $\delta_{ij} \Lambda^{ij}{}_{kl}(\mathbf{n}) = 0$, and
c) a projector, i.e. that $\Lambda^{ij}{}_{kl}(\mathbf{n}) \Lambda^{kl}{}_{mn}(\mathbf{n}) = \Lambda^{ij}{}_{mn}(\mathbf{n})$.

A3.2   [Hard!]
a) Verify that the Riemann tensor is given by equation (5.64) in a local inertial frame, and hence equation (A3.3) in the linearised approximation. You will need to start from the full expression, equation (5.47), and the formula for the Christoffel symbol equation (5.12).
b) Show that the Ricci tensor in the linearised approximation and Lorenz gauge is

$$R_{\nu\alpha} = -\frac{1}{2}\Box h_{\nu\alpha}.$$

c) Hence, show that the linearised Einstein equation for the trace-reversed metric is indeed equation (A3.7).

A3.3   Show that equation (A3.8) for the metric perturbation $h^{\mu\nu}$ follows from the equation (A3.7) for the trace-reversed metric perturbation $\bar{h}^{\mu\nu}$.

A3.4   Show that the amplitude of a gravitational wave $A_{\mu\nu}$ transforms under an infinitesimal coordinate transformation of the form $x^{\mu} \to x^{\mu} + \tilde{\xi}^{\mu} e^{ik \cdot x}$ as

$$A_{\mu\nu} \to A_{\mu\nu} + ik_{\mu}\tilde{\xi}_{\nu} + ik_{\nu}\tilde{\xi}_{\mu}.$$

Use this transformation to show that a transverse–traceless gravitational-wave amplitude can always be written as in equation (A3.13).

Advanced Topic A4

Gravitational Wave Sources and Detection

Prerequisites: Chapters 1 — 10, Advanced Topic A3

A century after their prediction, gravitational waves are finally directly detected · the first observed source was a merger of two black holes · such compact binaries emit a characteristic 'chirp' signal as they approach merger

One of the most impressive scientific results of the early 21st century is the first direct detection of gravitational waves, achieved by the Advanced Laser Interferometer Gravitational-Wave Observatory (abbreviated to AdvLIGO, or more commonly just LIGO). This experiment correlated signals received at two sites in the United States, at Hanford in Washington and Livingston in Louisiana, which are nearly 4000 km apart. Each observatory features two arms of length 4 km in an ultra-high vacuum state, through which high-intensity laser beams travel multiple times before being recombined to create an interference pattern. This gives an extraordinary sensitivity to changes in the arm length of about one part in 10^{21}.

In February 2016 came the announcement of the first direct detection of gravitational waves. The signal was received at the detectors in September 2015, just a few months before the 100th anniversary of General Relativity, and named GW150914. The source was identified as the merger of two black holes with masses of about 35 and 30 Solar masses, located at a distance of approximately 440 megaparsecs from us. The entire duration of the detectable signal, seen prominently by both detectors, was only 0.2 seconds.

This announcement marked the dawn of a new era of observational astronomy, using gravitational waves in place of light and probing entirely new physical regimes. The number of detections is rising rapidly as we write. Notable amongst them is GW170817, a merger of two neutron stars, which was also observed at

Introducing General Relativity, First Edition. Mark Hindmarsh and Andrew Liddle.
© 2022 John Wiley & Sons Ltd. Published 2022 by John Wiley & Sons Ltd.
Companion website: www.wiley.com/go/hindmarsh/introducingGR

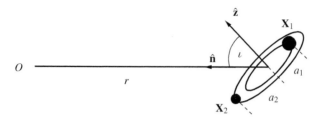

Figure A4.1 The geometry of a binary system, with observer located at O at a distance r from the centre of mass. The normal to the plane of the binary system $\hat{\mathbf{z}}$ is oriented at an angle ι to the direction of propagation of the gravitational waves $\mathbf{n}$.

various electromagnetic frequencies including gamma rays and optical light and shows that the propagation speed of gravitational waves is practically identical (to within one part in 10^{15}) of that of light. Astronomy from combining electromagnetic and gravitational-wave signals is called multi-messenger astronomy.

This Advanced Topic focuses on the physics of compact binary sources, their gravitational-wave signatures, and the detection of those signatures.

A4.1   Gravitational waves from compact binaries

In Advanced Topic A3, we found that the source for gravitational waves was the second time derivative of the quadrupole moment. Let us examine the quadrupole moment for a pair of point masses m_1 and m_2 in a circular orbit around their centre of mass, which we take to be the origin of the coordinates. We can also choose the orbit to be in the x–y plane. This will be our model for a pair of binary compact objects like white dwarfs, neutron stars, or black holes (see Figure A4.1).

If we make the assumption that the point masses are moving slowly compared with the speed of light, the mass density can be written

$$\frac{1}{c^2}T^{00} = m_1\delta^3(\mathbf{x}' - \mathbf{X}_1(t)) + m_2\delta^3(\mathbf{x}' - \mathbf{X}_2(t)),\qquad (A4.1)$$

where δ^3 is the three-dimensional Dirac delta function. The positions of the two masses $\mathbf{X}_1$, $\mathbf{X}_2$ are given by

$$\mathbf{X}_1(t) = a_1\left(\cos\phi(t)\hat{\mathbf{x}} + \sin\phi(t)\hat{\mathbf{y}}\right);\quad \mathbf{X}_2(t) = -a_2\left(\cos\phi(t)\hat{\mathbf{x}} + \sin\phi(t)\hat{\mathbf{y}}\right),$$
$$(A4.2)$$

where the azimuthal angle $\phi = \omega t$, and ω is the angular frequency. Hence, from equation (A3.29), the second moment of the mass distribution is

$$I^{ij} = I_{cm} \left[\hat{x}^i \hat{x}^j \cos^2 \phi + \hat{y}^i \hat{y}^j \sin^2 \phi + \left(\hat{x}^i \hat{y}^j + \hat{y}^i \hat{x}^j \right) \sin \phi \cos \phi \right], \qquad \text{(A4.3)}$$

where $I_{cm} = m_1 a_1^2 + m_2 a_2^2$ is the moment of inertia around the centre of mass. The second time derivative is therefore

$$\ddot{I}^{ij} = -2\omega^2 I_{cm} \left[\left(\hat{x}^i \hat{x}^j - \hat{y}^i \hat{y}^j \right) \cos(2\phi) + \left(\hat{x}^i \hat{y}^j + \hat{y}^i \hat{x}^j \right) \sin(2\phi) \right]. \qquad \text{(A4.4)}$$

We can see that this is traceless, as required for the quadrupole moment (A3.31). Note also that the waveform is transverse in the z direction, and we can identify the polarisation basis tensors (A3.15) already in the solution. Hence, the polarisation components in the direction $\mathbf{n} = \hat{\mathbf{z}}$ are equal in amplitude and $\pi/2$ out of phase,

$$\bar{h}_+^{ij}(\hat{\mathbf{z}}) = \frac{4GI_{cm}\omega^2}{rc^4} \cos(2\omega t); \quad \bar{h}_\times^{ij}(\hat{\mathbf{z}}) = \frac{4GI_{cm}\omega^2}{rc^4} \sin(2\omega t), \qquad \text{(A4.5)}$$

which is the gravitational-wave version of circular polarisation. The frequency of the gravitational waves is twice the orbital frequency of the binary system.

One can find the polarisations in other directions using the projector $\Lambda^{ij}{}_{kl}(\mathbf{n})$. The resulting metric perturbation in a gravitational wave in a general direction $\mathbf{n}$ oriented at angle ι with respect to the orbital plane is

$$\bar{h}_{TT}^{ij}(\mathbf{n}) = \frac{4GI_{cm}\omega^2}{rc^4} \left[\frac{1}{2}(1 + \cos^2 \iota) \cos(2\omega t) \varepsilon_+^{ij}(\mathbf{n}) + \cos \iota \sin(2\omega t) \varepsilon_\times^{ij}(\mathbf{n}) \right].$$
$$\text{(A4.6)}$$

It is a good exercise to verify this formula (see Problem A4.2). Note that the $\times$ polarisation vanishes for gravitational waves emitted in the x–y plane.

We are now in a position to estimate the gravitational-wave amplitude from a binary system of compact objects, such as a binary black hole. We first write the total mass as $M = m_1 + m_2$ and estimate the moment of inertia as Ml^2, where $l = a_1 + a_2$ is the distance between the objects. The order of magnitude of the components of the wave amplitude is

$$\bar{h} \sim \frac{GM}{rc^2} \frac{v^2}{c^2}, \qquad \text{(A4.7)}$$

where $v = l\omega/2$ is the approximate orbital speed. Clearly the strongest signal comes from the fastest orbits. For a pair of $30M_\odot$ black holes at a distance from us of 400 megaparsecs (the approximate masses and distances of the first LIGO

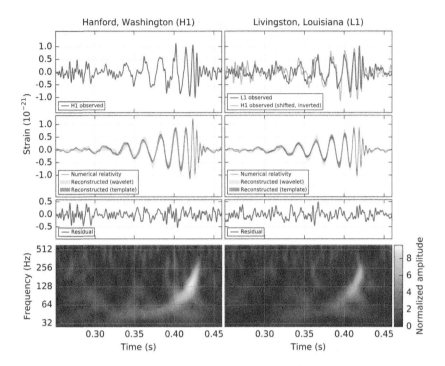

Figure A4.2 LIGO waveforms of the first direct gravitational wave detection, GW150914. (From *Phys. Rev. Lett.*, **116**, 061102 (2016), reproduced under Creative Commons license.)

event GW150914), we find

$$\bar{h} \sim 3 \times 10^{-21} \frac{v^2}{c^2}. \tag{A4.8}$$

The strongest signal comes just before the black holes collide, where they are moving close to the speed of light. The order of magnitude estimate can be compared with the actual signal observed by LIGO in Figure A4.2, which is a little lower. Our estimate did not make allowance for the orientation of the binary or the detector; furthermore the weak-field calculation assumed slow motion ($v \ll c$), and the black holes were in fact travelling at about $v \simeq 0.5c$ when they merged.

It is also instructive to use Keplerian dynamics to estimate that $v^2 \simeq 2GM/l$, in which case

$$\bar{h} \sim \frac{r_S}{r} \frac{r_S}{l}, \tag{A4.9}$$

where r_S is the Schwarzschild radius. Hence, the gravitational-wave amplitude is roughly the product of the gravitational potential near the binary with the gravitational potential where the radiation is being observed.

A4.2   The energy in gravitational waves

A notable feature of the signal is that the wave crests get closer together: as the black holes lose energy to gravitational radiation, they move closer together, and hence their orbital period decreases. This process is known as binary inspiral.

In order to estimate the rate at which the period changes, we need to know the energy carried away by the gravitational waves. The notion of energy carried by the gravitational field is not an obvious one. Energy belongs to the energy–momentum tensor of the matter, and the gravitational field is described by the Einstein tensor $G^{\mu\nu}$. They are related by the Einstein equation, $G^{\mu\nu} = \kappa T^{\mu\nu}$. However, by looking at the higher-order terms in the weak-field expansion of the Einstein tensor, one can formally assign the $O(h^2)$ terms to the energy–momentum, and obtain an expression which looks like an energy–momentum associated with the gravitational waves. The flux (power per unit area) is

$$t^0_i = \frac{c^4}{32\pi G} \left\langle \dot{\bar{h}}_{\mathrm{TT}\alpha\beta} \bar{h}^{\alpha\beta}_{\mathrm{TT},i} \right\rangle , \qquad (A4.10)$$

where the angle brackets denote an average over at least one period of the wave. The luminosity of a source is the power radiated through a large enclosing sphere with unit normal n^i,

$$L = \int d\Omega\, r^2 n^i t^0_i = \frac{1}{2} \frac{G}{c^5} \int \frac{d\Omega}{4\pi} \left\langle \dddot{I}^{ij}_{\mathrm{TT}} \dddot{I}_{\mathrm{TT}ij} \right\rangle . \qquad (A4.11)$$

The direction of the projector **n** can be integrated over, and a lengthy calculation[1] shows that

$$L = \frac{1}{5} \frac{G}{c^5} \left\langle \dddot{I}^{ij} \dddot{I}_{ij} \right\rangle . \qquad (A4.12)$$

The luminosity of a gravitational wave source is therefore proportional to the square of the third time derivative of the quadrupole moment.

---

[1] Details of the integration, and of the construction of the energy–momentum tensor, can be found in Schutz's book.

A4.3   Binary inspiral

We can now study how the orbital period, or equivalently the radius of the orbit, changes as a result of the energy loss to radiation. We know that we will need to take time derivatives of the quadrupole moment, so we start with the second moment of the mass distribution

$$\dddot{I}^{kl} = 2\omega^3 I_{cm} \left[\left(\hat{x}^i \hat{x}^j - \hat{y}^i \hat{y}^j \right) \sin(2\phi) - \left(\hat{x}^i \hat{y}^j + \hat{y}^i \hat{x}^j \right) \cos(2\phi) \right] . \qquad (A4.13)$$

Hence, using the definition of the quadrupole moment (A3.31), we find that

$$\left\langle \dddot{I}^{ij} \dddot{I}_{ij} \right\rangle = 32 \omega^6 I_{cm}^2 , \qquad (A4.14)$$

and so the luminosity in gravitational waves is

$$L = \frac{32}{5} \frac{G}{c^5} I_{cm}^2 \omega^6 . \qquad (A4.15)$$

We now turn to the total energy of the binary system, which in the Newtonian approximation is

$$E = \frac{1}{2} \mu l^2 \omega^2 - \frac{G\mu M}{l} , \qquad (A4.16)$$

where $\mu = m_1 m_2 / M$ is the reduced mass and $l = a_1 + a_2$ is the separation between pair. Using Kepler's Law, $\omega^2 l^3 = GM$, we can write the energy as

$$E = -\frac{G\mu M}{2l} = -\frac{1}{2} \mu c^2 \left(\frac{GM\omega}{c^3} \right)^{2/3} , \qquad (A4.17)$$

where the factor in the brackets is a convenient dimensionless combination. The total energy is negative, as it should be for a bound system. The luminosity can also be re-expressed using Kepler's Law, and noting that $I_{cm} = \mu l^2$,

$$L = \frac{32}{5} \omega \mu c^2 \left(\frac{G\mu\omega}{c^3} \right) \left(\frac{GM\omega}{c^3} \right)^{4/3} . \qquad (A4.18)$$

If the main source of energy loss is gravitational radiation, the rate of change of the energy can be equated with the luminosity in gravitational waves, to derive an equation for the time variation of the angular frequency ω. Defining the *chirp mass*

$$\mathcal{M} = \mu^{3/5} M^{2/5} = \frac{(m_1 m_2)^{3/5}}{(m_1 + m_2)^{1/5}}, \tag{A4.19}$$

we find

$$\frac{\dot{\omega}}{\omega} = \frac{96}{5} \omega \left(\frac{G \mathcal{M} \omega}{c^3} \right)^{5/3}. \tag{A4.20}$$

This evolution of the binary system can also be expressed in terms of the orbital period P,

$$\dot{P} = -\frac{192\pi}{5} \left(\frac{2\pi G \mathcal{M}}{c^3 P} \right)^{5/3}. \tag{A4.21}$$

We therefore learn that an observation of the gravitational wave frequency $f = 2\omega$ and its rate of change is sufficient to give a measurement of the chirp mass.

The first system to be discovered for which gravitational radiation losses were important was a binary consisting of pulsar and another neutron star, which we have already encountered in Chapter 10. The eccentricity of the orbit is quite important, as the velocities of the neutron stars are much higher than average when they are close together, and the gravitational-wave luminosity is very sensitive to the velocities of the objects involved. The net result is to multiply the right-hand side in equation (A4.21) by a correction factor

$$k(\varepsilon) = \left(1 + \frac{73}{24} \varepsilon^2 + \frac{37}{96} \varepsilon^4 \right) \left(1 - \varepsilon^2 \right)^{-7/2}. \tag{A4.22}$$

When the objects in a binary are closer together, we can hope to see more of the evolution of the orbital frequency. Equation (A4.20) can be integrated once to give its time dependence, and again to give the time-dependence of the orbital phase $\phi(t)$. One finds that the frequency goes to infinity at a finite time, which we interpret as the coalescence of the binary, and denote t_{co}. Writing the solution in terms of the gravitational-wave frequency $f = 2/P$ and a constant with dimensions of time $T_{ch} = 5G \mathcal{M}/c^3$ gives

$$f(t) = \frac{5}{8\pi T_{ch}} \left(\frac{T_{ch}}{t_{co} - t} \right)^{3/8}. \tag{A4.23}$$

The amplitude of the gravitational wave also increases as the orbital frequency ω increases. Substituting $\omega = f(t)\pi$ into the order-of-magnitude expression for the

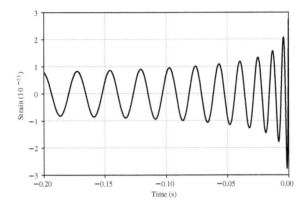

Figure A4.3 Estimate of the strain amplitude at the Earth of the gravitational wave emitted by a binary black hole, as a function of time before coalescence, as predicted by the weak field analysis of this section. The strain amplitude is taken from the right-hand side of equation (A4.24), using the central values of the black hole masses involved in GW150914, $m_1 = 35 M_\odot$ and $m_2 = 30 M_\odot$, and the central value of the distance of the system 410 megaparsecs.

wave amplitude, we find

$$\bar{h} \sim \frac{G\mathcal{M}}{rc^2} \left(\frac{T_{\text{ch}}}{t_{\text{co}} - t} \right)^{1/4}. \tag{A4.24}$$

This characteristic increase in frequency and amplitude is known as a **chirp**. Figure A4.3 shows the predicted shape of the chirp, which can be compared to the data shown in Figure A4.2. The agreement in the early stages is striking. A detailed comparison of the waveforms through the merger requires numerical solutions of the Einstein equations.

Only the chirp mass appears in equation (A4.21), and no other combination of the masses in the system. Therefore, given that the chirp mass can be inferred just from the gravitational wave period and its rate of change, the amplitude of the wave gives an estimate of the distance of the source. There is a significant uncertainty, as it is difficult to untangle the angle of the orbital plane to the direction of propagation.

The individual masses of the merging objects can be recovered from more accurate calculations for the gravitational binding energy E and the gravitational-wave luminosity L, going beyond the approximations we have made so far, which stem from the assumption that the motion of the objects is slow compared with light. The expressions (A4.17) and (A4.18) can be corrected by higher order terms in an expansion in $(GM\omega/c^3)^{1/3}$, which is the orbital speed in units of c in the non-relativistic approximation. It turns out that that the time dependence of the orbital phase of the binary $\phi(t)$ can be expressed in terms of an expansion in powers of the symmetric mass ratio $\nu = \mu/M$ and the dimensionless variable $\tau = \left[5c^3\nu(t_{co} - t)/\pi GM\right]^{-1/8}$. The first terms in the expansion are[2]

$$\phi(t) = -\frac{1}{\nu\tau^5}\left[1 + \left(\frac{3715}{8064} + \frac{55}{96}\nu\right)\tau^2\right].$$ (A4.25)

Hence, with good data and careful fitting, the total mass M and the symmetric mass ratio ν can be separately determined, and hence the masses of objects involved. For GW150914, the masses were found to be

$$m_1 = \left(35^{+5}_{-3}\right)M_\odot; \quad m_2 = \left(30^{+3}_{-4}\right)M_\odot,$$ (A4.26)

although the distance could only be pinned down to 440^{+160}_{-180} megaparsecs due to the degeneracy with the angle of the orbital plane. That it was possible to determine the masses so accurately, for the first gravitational event to be observed, is a piece of remarkable good fortune.

A4.4   Detecting gravitational waves

A4.4.1   Laser interferometers

The two instruments which detected the gravitational waves from GW150914 were laser interferometers. The basic design principle of an interferometric gravitational wave detector is very simple: a gravitational wave passing through the interferometer causes a relative change in the length of the arms, and so the relative phase of the two light beams at the output is shifted. If the arm lengths are constructed so that there is normally destructive interference at the output, a passing gravitational wave will induce a signal in a photodetector placed there (see the sketch of a Michelson interferometer in Figure A4.4).

Let us calculate how a gravitational wave with metric $h_{ij}(\mathbf{x}, t)$ affects the phase of the light travelling along one of the arms. We drop 'TT' notation on the understanding that h_{ij} is transverse and traceless in this section. We choose the arm to lie along the x-axis, and have length L, with the beam splitter located at $x = 0$. We can calculate the phase change from the change in the light travel time to go from

---

[2]Details of this expansion can be found in B.S. Sathyaprakash and B.F. Schutz, *Living Rev. Rel.* **12**, 2 (2009).

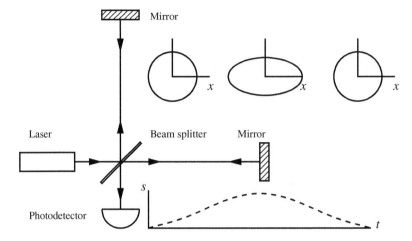

Figure A4.4 Diagram of a gravitational wave detector based on a Michelson interferometer. Normally, the lengths of the arms are arranged so that the light at the output undergoes destructive interference. The mirrors and the beam splitter are suspended and free to move. A gravitational wave passing through changes the relative lengths of the two arms, causing a signal $s(t)$ at the photodetector.

the beam splitter to the mirror and back again. Light travels on null geodesics of the metric, and so the trajectory $x(\lambda)$, $t(\lambda)$ (where λ is an affine parameter) satisfies

$$ds^2 = -c^2 dt^2 + dx^2(1+h_{xx}(x(\lambda),t(\lambda))) = 0. \qquad (A4.27)$$

Hence, the time taken to travel from the beam splitter to the mirror and back again along the x arm is

$$T_x = \int \frac{dx}{d\lambda}(1+h_{xx}(x(\lambda),t(\lambda)))^{1/2} d\lambda. \qquad (A4.28)$$

The metric perturbation is so small that we can Taylor expand the square root, and use the unperturbed geodesic to evaluate h_{xx}. The errors are of order h^2, and therefore negligible.

The unperturbed geodesic is

$$t = (\lambda - 2L/c) + t_a; \quad x = \begin{cases} c\lambda & 0 \le c\lambda < L \\ 2L - c\lambda & L \le c\lambda \le 2L \end{cases}, \qquad (A4.29)$$

where t_a is the arrival time at the beam splitter. Hence, $T_h = 2L/c + \delta T$ where

$$\delta T_x = \frac{1}{2} \int_0^{2L} h_{xx}(x(\lambda), t(\lambda)) d\lambda. \tag{A4.30}$$

The metric perturbation corresponding to a gravitational wave is

$$h_{xx} = A_{xx} e^{-i\omega t + i\mathbf{k} \cdot \mathbf{x}} + \text{c.c.}. \tag{A4.31}$$

For simplicity, let us suppose that the gravitational wave is normally incident on the plane of the interferometer. Then, it is not hard to show that

$$\delta T_x = \frac{1}{\omega} A_{xx} e^{i\omega (L/c - t_a)} \sin(\omega L/c) + \text{c.c.}. \tag{A4.32}$$

If the wavelength of the gravitational wave is much longer than the arm of the interferometer, as is the case for current gravitational wave detectors,

$$\delta T_x \simeq \frac{L}{c} A_{xx} e^{-i\omega t_a} + \text{c.c.}. \tag{A4.33}$$

We can do the same calculation for the y arm for δT_y. The difference in arrival times at the beam splitter, where the light beams interfere, is $\delta T = \delta T_x - \delta T_y$. The phase difference Φ between the beams for light of frequency f_γ (not to be confused with the Newtonian gravitational potential) is therefore

$$\Phi(t_a) = \frac{2\pi f_\gamma L}{c} (|A_{xx}| - |A_{yy}|) \cos(\omega t_a + \varphi), \tag{A4.34}$$

where φ is the phase of the gravitational wave.

We can equally well think of the time difference as having been caused by a fractional change in the length of the arms δL,

$$\delta L_x \simeq L |A_{xx}| \cos(\omega t_a + \varphi). \tag{A4.35}$$

Gravitational wave signals are often described in terms of the strain $h = \delta L_x / L$, which is of magnitude $|A_{xx}|$, and hence equal to the magnitude of the metric perturbation projected along the arms.

Building a real interferometric gravitational wave detector capable of detecting strains of order 10^{-21}, such as LIGO, is an enormous technical challenge. One wants to have as long arms as possible, so that the mirrors move a greater distance. One also wants to have as an intense a laser as possible so that the small amount of light which the gravitational wave lets through is detectable, and one

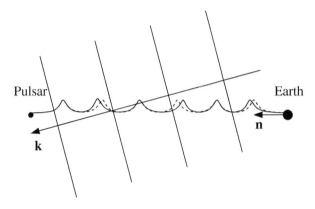

Figure A4.5   Pulsar timing as a detection method for gravitational waves. A pulsar in direction **n** emits very regular pulses of radio waves which travel towards Earth, where they are detected by radio telescopes. An intervening gravitational wave with wave vector **k** causes a small displacement of the arrival times of the pulses, indicated by the dashed line. See Problem A4.1 for the calculation of the arrival time displacement.

needs to isolate the system from other sources of vibration, such as earthquakes and the thermal vibrations of the mirrors.

The laser power is boosted by using evacuated Fabry–Pérot cavities instead of a simple Michelson interferometer. The light bounces back and forth between the mirrors in the cavity many times, which also multiplies the effective arm length. The mirrors of the cavity are isolated from seismic noise by an elaborate suspension system. Thermal noise is reduced by mounting the mirrors on very strong and stiff wires whose thermal excitations occur at controllable frequencies. The next generation of gravitational wave experiments, such as the Kamioka Gravitational Wave Detector (KAGRA) in Japan, cool the mirrors and are located underground.

A4.4.2   Pulsar timing

Another method of detecting gravitational waves exploits pulsars as very accurate clocks distributed throughout the galaxy. Pulsars are rapidly rotating neutron stars which emit narrow beams of intense radiation from their magnetic poles. If the beams happen to cross the Earth, we see very regular radio pulses, separated by as little as a few milliseconds. If a gravitational wave passes between the Earth and the pulsar, the arrival time of the pulses will be displaced (see Figure A4.5 and Problem A4.1 for a calculation of the arrival time displacement).

By monitoring many pulsars, one can average out uncertainties in the expected pulse arrival time due to, for example, propagation of the radio waves through the interstellar medium. At the time of writing, three such projects are running, NANOGrav (the North American Nanohertz Observatory for Gravitational waves), PPTA (the Parkes Pulsar Timing Array) based in Australia, and EPTA (the European Pulsar Timing Array). Such experiments are most sensitive to gravitational waves whose period is the same order of magnitude as a the length of the observations, which is about 10 years for many of the pulsars concerned. Excitingly, in 2020 NANOGrav reported a signal consistent with being due to gravitational waves. More data is awaited to determine whether the variation in the pulse arrival times has the characteristic angular distribution around the sky expected from gravitational waves. If confirmed, the most likely source would be mergers of two black holes of masses around $10^9 \, M_\odot$, brought together by the collision and merger of distant galaxies.

Many more pulsars will be monitored by a large radio telescope system called the Square Kilometre Array (SKA), currently being built in South Africa and Australia, and due to start observations around 2027. This array of telescopes is predicted to be one hundred times more sensitive than the current projects, and will surpass them as a gravitational-wave detector within a few years.

A4.4.3 Interferometers in space

Further in the future, gravitational wave interferometers will move into space to eliminate seismic noise altogether, and to allow arm lengths of millions of kilometres so that lower frequency sources can be detected. The Laser Interferometer Space Antenna (LISA) is scheduled to be launched by 2034, and will consist of three spacecraft orbiting the Sun behind the Earth, monitoring changes in each other's relative distance with pairs of laser 'links' (see Figure A4.6). It will be most sensitive to gravitational waves with frequency around 1 mHz.

As well as seeing the early stages of the kind of binary black holes whose merger was seen by LIGO, LISA will be able to detect the inspiral of more massive binary black holes, which are known to exist in the cores of galaxies. It will be sensitive to the merger of black holes with masses in the range $10^4 < M/M_\odot < 10^6$ at redshifts of up to about 10. It will also have the sensitivity to detect violent events in the very early Universe occurring at around a few picoseconds after the Big Bang, where thermal energies and densities were so high that the physical model describing the contents of the universe is not well understood. This was the era when the Higgs field turned on and elementary particles gained masses. If the Standard Model of Particle Physics describes this process well, then it would have been a smooth transition. However, in many extensions of the Standard Model, the turning-on of the Higgs field would have been sudden and violent, rather like lots of explosions all over the Universe. The colliding shock waves from these explosions would have generated gravitational waves of about the right frequency

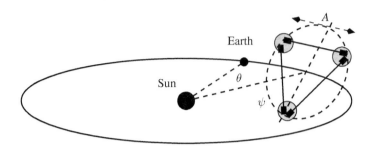

Figure A4.6 The LISA orbit. LISA will consist of three spacecraft in an equilateral triangle, separated by $A = 2.5$ million kilometres, orbiting around the Sun about $\theta = 20°$ behind the Earth. Their orbits will inclined so that the triangle rotates in a plane oriented at $\psi = 60°$ to the plane of the Earth's orbit. Each pair of satellite will be linked by two lasers, which will allow very accurate interferometric determination of their relative distance. A gravitational wave passing through the constellation of satellites will cause the satellites to approach and recede from one another in a characteristic way.

to be detected by LISA. Besides investigating enormous black holes at high red-shifts, LISA will be a particle physics experiment, with the ability to complement the discoveries made at the Large Hadron Collider in its high luminosity phase.

   Other space-based gravitational wave observatories are planned. The most advanced are the Chinese projects Taiji (which will be very similar in size and sensitivity to LISA) and TianQin, which will have three spacecraft orbiting the Earth inside the Moon's orbit. Both are due for launch in the 2030s.

Problems

A4.1   [Hard!] Radio pulses are emitted by a pulsar at distance L in direction $\mathbf{n}$, and travel through a gravitational wave of amplitude A_{ij} and wavenumber $\mathbf{k}$ (see Figure A4.5). Show that the arrival times of the pulses are displaced by

$$\delta T(t_a) = \frac{L}{c}|n^i n^j A_{ij}| \frac{2\sin(x/2)}{x} \cos\left(\frac{x}{2} + \phi - \omega t_a\right),$$

where $x = \omega(1 + \hat{\mathbf{k}} \cdot \mathbf{n})L/c$, and φ is a phase angle.

A4.2   [Hard!] Verify equation (A4.6) for the metric perturbation in an arbitrary direction from a binary system.
[Hint: the two key formulae are equations (A3.35) and (A4.4).]

Bibliography

A vast array of books on General Relativity and gravity exists, catering for all levels of technical sophistication. Here is a far-from-complete selection. We consider this book to be Introductory Undergraduate.

Background

'Einstein's Masterwork: 1915 and the General Theory of Relativity', J. Gribbin, Icon Books, 2016:
An authoritative popular-level introduction to the topic by one of the very best science writers.

'The Ascent of Gravity', M. Chown, W&N, 2017:
'The Perfect Theory', P. G. Ferreira, Abacus, 2015:
Two general accounts of the development of the modern theory of gravitation, explaining the key concepts underlying this book in non-mathematical terms.

'Subtle is the Lord', A. Pais, Oxford University Press, 2005 (first published 1982):
The definitive autobiography of Albert Einstein, which also explains his science in quite some technical detail.

Einstein's original papers, along with English translations, are available as *The Collected Papers of Albert Einstein* at
`http://einsteinpapers.press.princeton.edu`.

Undergraduate

'A First Course in General Relativity', B. Schutz, Cambridge University Press, 2009 (2nd ed.):
An excellent introductory text, operating at a somewhat higher level of mathemat-

Introducing General Relativity, First Edition. Mark Hindmarsh and Andrew Liddle.
© 2022 John Wiley & Sons Ltd. Published 2022 by John Wiley & Sons Ltd.
Companion website: www.wiley.com/go/hindmarsh/introducingGR

ical sophistication than the present book.

'Relativity: The Special and General Theory', A. Einstein, various publishers, translated from German (first published 1916):
Why not have a go at learning from the great scientist himself!

'Special Relativity', A. P. French, CRC Press, 1968:
'Introduction to the Relativity Principle', G. Barton, John Wiley & Sons, 1999:
If you need to brush up on your special relativity before embarking on this book, try these two, the former a classic and much-used text and the latter taking a more modern viewpoint.

'Spacetime and Geometry', S. Carroll, Pearson, 2013:
A beautifully-written book which brings out the conceptual elegance of general relativity.

'General Relativity: An Introduction for Physicists', M. P. Hobson, G. P. Efstathiou, and A. N. Lasenby, Cambridge University Press, 2006:
While the emphasis is on applications to physics and astrophysics, this book is also quite demanding on the mathematical side.

Postgraduate/researcher

'Gravitation and Cosmology', S. Weinberg, John Wiley & Sons, 1972:
A landmark and definitive text, renowned for the clarity of its presentation.

'Gravitation', C. W. Misner, K. Thorne, and J. A. Wheeler, W. H. Freeman, 1973:
Written in a unique style that is by turns inspiring, challenging, and maddening over its thousand-plus large-format pages, this is the one textbook that should be read by anyone wishing to call themselves a general relativist.

'General Relativity', R. M. Wald, University of Chicago Press, 1984:
Comprehensive and mathematically rigorous, this book is often the first port of call for an active relativity researcher in search of ways of addressing their problems.

'Problem Book in Relativity and Gravitation', A. Lightman, W. H. Press, R. H. Price, and S. A. Teukolsky, Princeton University Press, 2017 (first published 1974):
More General Relativity problems than could be solved in a human lifetime, for those who like to learn by doing rather than just reading.

'Theory and Experiment in Gravitational Physics', C. Will, Cambridge University Press, 1993:
The definitive book on experimental tests of gravity, by the recognised leading authority on the topic. See also his *Living Reviews* article given below for more up-to-date information.

In addition to the above textbooks, an invaluable source for researchers into relativity is the on-line journal 'Living Reviews in Relativity', a compilation which contains a large number of frequently-updated articles on a wide range of topics relating to gravitation and General Relativity. We have selected two of particular relevance:
'Physics, Astrophysics and Cosmology with Gravitational Waves', B. S. Sathyaprakash and B. F. Schutz, Living Rev. Rel. **12** (2009) 2.
'The Confrontation between General Relativity and Experiment', Clifford M. Will, Living Rev. Rel. **17** (2014) 4.

Finally, we mention the very useful on-line article 'Catalogue of spacetimes' by T. Mueller and F. Grave, available at `http://arXiv.org/abs/0904.4184`, which gives explicit formulae for the Christoffel symbols and curvature tensors of a large range of space–times including 8 different coordinate systems for the Schwarzschild geometry. Excellent for checking your own evaluations!

Answers to Selected Problems

Here we provide solutions, some in detail and some in outline, for many of the problems posed at the end of each chapter. A star next to the problem number indicates that there is a solution here. A complete set of solutions to all problems is available online to instructors who register at the book's companion website.

Chapter 2

2.1 The Lorentz transformation along the x-axis is

$$ct' = \gamma \left(ct - \frac{v}{c}x \right); \quad x' = \gamma(x - vt).$$

Start with

$$s^2 = -c^2 t'^2 + (x')^2 + (y')^2 + (z')^2,$$

and substitute in the Lorentz transformation. Expand out the brackets, remembering $\gamma \equiv 1/\sqrt{1 - (v/c)^2}$, and this will simplify to

$$s^2 = -c^2 t^2 + x^2 + y^2 + z^2,$$

which is the invariance we want (i.e. the expression is the same in both coordinate systems).

For a general Lorentz transformation, we can always rotate the coordinates so the transformation is along the x-axis and then use the above result.

2.2 There are two possible methods: either use time dilation for the fast-moving muon in our rest frame, or use Lorentz–Fitzgerald contraction of the atmosphere in the rest-frame of the muon. In the former case, the calculation can be simplified with the legitimate approximation that in working out the distance it travels, the muon can be taken to go at the speed of light in our rest frame.

Introducing General Relativity, First Edition. Mark Hindmarsh and Andrew Liddle.
© 2022 John Wiley & Sons Ltd. Published 2022 by John Wiley & Sons Ltd.
Companion website: www.wiley.com/go/hindmarsh/introducingGR

First note that the half-life is indeed the time needed for the muon to have a half-chance of surviving. Let's use Lorentz–Fitzgerald contraction. Seen by the muon, the thickness of the atmosphere is $30\,\text{km} \times \sqrt{1 - v^2/c^2}$. It has 2.2×10^{-6} seconds of existence, during which the Earth travels a distance $2.2 \times 10^{-6}v$. The minimum velocity is therefore given by (being careful to convert kilometres to metres)

$$2.2 \times 10^{-6}v = 30000\sqrt{1 - v^2/c^2}.$$

Square both sides and substitute in for c on the left to get

$$4.84 \times 10^{-4}v^2/c^2 = 1 - v^2/c^2,$$

and so $v = 0.9998c$. [Note: it is best to leave the answer as this rather than substituting in for c.]

2.4

a) This amounts to demonstrating the not-at-all obvious trigonometric identity

$$\tanh(a+b) = \frac{\tanh a + \tanh b}{1 + \tanh a \tanh b},$$

which is the hyperbolic equivalent of the slightly more familiar identity $\tan(a+b) = (\tan a + \tan b)/(1 - \tan a \tan b)$.

b) First show that $\gamma = \cosh W$, and then substitution into equation (2.7) immediately gives the answer.

c) Carry out the same process as Problem 2.1 using this new version of the Lorentz transformation.

Chapter 3

3.1

a) We substitute the Lorentz transformation

$$x'^\rho = \sum_{\sigma=0}^{3} \Lambda^\rho_{\ \sigma} x^\sigma,$$

into the space–time interval to get

$$ds'^2 = \sum_{\mu\nu} \eta_{\mu\nu} dx'^\mu dx'^\nu = \sum_{\mu\nu\rho\sigma} \eta_{\mu\nu} \Lambda^\mu_{\ \rho} dx^\rho \Lambda^\nu_{\ \sigma} dx^\sigma = \sum_{\rho\sigma} \eta_{\rho\sigma} dx^\rho dx^\sigma,$$

where the last equality follows from invariance of the interval.

The final equality has to hold true for all choices of dx^ρ and dx^σ, and this can only be true if the required condition holds.

b) In matrix multiplication, the summation indices pair a row in the first matrix with a column in the second matrix, and we write the row index first and the column index second. Inspecting the equation, we see that if we interpret $\eta_{\mu\nu}$ as a matrix, with μ labelling rows and ν labelling columns, then

$$\sum_{\mu\nu} \eta_{\mu\nu} \Lambda^\mu{}_\rho \Lambda^\nu{}_\sigma = \sum_{\mu\nu} \Lambda^\mu{}_\rho \eta_{\mu\nu} \Lambda^\nu{}_\sigma = \sum_{\mu\nu} (\Lambda^T)_\rho{}^\mu \eta_{\mu\nu} \Lambda^\nu{}_\sigma \,.$$

This is now the component form of the matrix equation $\Lambda^T \eta \Lambda = \eta$. Note that taking the transpose does not affect the level of the index.

c) The Lorentz transformation in matrix form is $a' = \Lambda a$, hence $a'^T = a^T \Lambda^T$. So

$$a' \cdot b' = a'^T \eta\, b' = a^T \Lambda^T \eta\, \Lambda b = a^T \eta\, b = a \cdot b \,.$$

3.4

a) You can write out the matrices and do the sum down the diagonal if you want, but you can also use the definition $\eta_{\mu\nu} \eta^{\nu\rho} = \delta^\rho_\mu$ with $\mu = \rho$. This gives

$$\eta_{\mu\nu} \eta^{\mu\nu} = \delta^\mu_\mu = 4 \,,$$

where the last follows because we have to do a sum over all four possible values of μ.

b) $\eta_{\mu\nu} \eta^{\mu\rho} \eta^{\nu\sigma} = \delta^\rho_\nu \eta^{\nu\sigma} = \eta^{\rho\sigma}$.

3.9

3.9 By definition, $\partial^\mu = \eta^{\mu\nu} \partial_\nu$, and the transformation law for the lower-index derivative is $\partial'_\nu = \Lambda_\nu{}^\rho \partial_\rho$. So

$$\partial^\mu \to \partial'^\mu = \eta^{\mu\nu} \partial'_\nu = \eta^{\mu\nu} \Lambda_\nu{}^\rho \partial_\rho = \eta^{\mu\nu} \Lambda_\nu{}^\rho \eta_{\rho\sigma} \partial^\sigma = \Lambda^\mu{}_\sigma \partial^\sigma \,,$$

which is the required answer.

3.10 First, we take the scalar product of the velocity and force 4-vectors,

$$u_\mu f^\mu = -q F^{\mu\nu} u_\mu u_\nu \,.$$

Then we note that, due to the Einstein summation convention, the labelling of repeated indices is arbitrary, and we can exchange the labels μ and ν, so

$$F^{\mu\nu} u_\mu u_\nu = F^{\nu\mu} u_\nu u_\mu .$$

Now we use the antisymmetry of the field-strength tensor, and the fact that we can write the product of the velocity components in any order,

$$F^{\nu\mu} u_\nu u_\mu = -F^{\mu\nu} u_\mu u_\nu .$$

Hence, the right-hand side of the first equation above is equal to minus itself, and is therefore zero,

$$u_\mu f^\mu = -qF^{\mu\nu} u_\mu u_\nu = qF^{\mu\nu} u_\mu u_\nu = 0 .$$

3.12   Here it helps to think of the Lorentz transformation as matrix multiplication, with $F \rightarrow F' = \Lambda F \Lambda^T$. To make the notation more compact, we will write $\beta^i = v^i/c$ and $e^i = E^i/c$, and write out the components of the matrix Λ only when necessary. Hence

$$
F' = \Lambda
\begin{pmatrix}
0 & e^1 & e^2 & e^3 \\
-e^1 & 0 & -B^3 & B^2 \\
-e^2 & B^3 & 0 & -B^1 \\
-e^3 & -B^2 & B^1 & 0
\end{pmatrix}
\begin{pmatrix}
\gamma & -\gamma\beta & 0 & 0 \\
-\gamma\beta & \gamma & 0 & 0 \\
0 & 0 & 1 & 0 \\
0 & 0 & 0 & 1
\end{pmatrix},
$$

$$
=
\begin{pmatrix}
\gamma & -\gamma\beta & 0 & 0 \\
-\gamma\beta & \gamma & 0 & 0 \\
0 & 0 & 1 & 0 \\
0 & 0 & 0 & 1
\end{pmatrix}
\begin{pmatrix}
-\gamma\beta e^1 & \gamma e^1 & e^2 & e^3 \\
-\gamma e^1 & \gamma\beta e^1 & -B^3 & B^2 \\
-\gamma e^2 - \gamma\beta B^3 & \gamma\beta e^2 + \gamma B^3 & 0 & -B^1 \\
-\gamma e^3 + \gamma\beta B^3 & \gamma\beta e^3 - \gamma B^2 & B^1 & 0
\end{pmatrix},
$$

$$
=
\begin{pmatrix}
0 & e^1 & \gamma(e^2 + \beta B^3) & \gamma(e^3 - \beta B^2) \\
 & 0 & -\gamma(B^3 + \beta e^2) & \gamma(B^2 - \beta e^3) \\
 & & 0 & -B^1 \\
 & & & 0
\end{pmatrix}.
$$

The other elements follow by antisymmetry. The transformation laws can be read off easily.

Chapter 4

4.1

   a) The gravitational potential around a spherical body is $\Phi = -GM/r$. As the potential at infinity is $\Phi = 0$, the gravitational redshift is

$$f(\infty) = f_{em}\left(1 - \frac{GM}{rc^2}\right).$$

b) $\Delta f/f_{em} = 2.5 \times 10^{-4}$. The corresponding velocity is $76\mathrm{km\,s^{-1}}$.
c) Zero. For the Earth, $GM/c^2 = 4.4$ mm. This result can't be trusted because the ratio $\Phi/c^2 = 1$, violating the assumption that the field is weak.

4.6

a) You will have done this in your mathematical methods courses, see e.g. Arfken, Weber, and Harris, *Mathematical Methods for Physicists*.
b) Differentiation of the relation $\rho = a\sin\theta$ gives $d\rho = a\cos\theta\,d\theta$ and hence $d\rho = (a^2 - \rho^2)^{-1/2}d\theta$.
c) From your mathematical methods courses you will also know that the Laplacian in spherical polar coordinates is

$$\nabla^2 = \frac{1}{r^2}\frac{\partial}{\partial r}\left(r^2\frac{\partial}{\partial r}\right) + \frac{1}{r^2\sin\theta}\frac{\partial}{\partial\theta}\left(\sin\theta\frac{\partial}{\partial\theta}\right) + \frac{1}{r^2\sin^2\theta}\frac{\partial^2}{\partial\phi^2}.$$

(You should remind yourself how this is proved.) At constant $r = a$, we can drop the differentiation with respect to the radius. The azimuthal angles is unchanged by the transformation, so

$$\frac{1}{a^2\sin^2\theta}\frac{\partial^2}{\partial\phi^2} = \frac{1}{\rho^2}\frac{\partial^2}{\partial\phi^2}.$$

We then just have to work out

$$\frac{\partial}{\partial\theta} = \frac{\partial\rho}{\partial\theta}\frac{\partial}{\partial\rho} = a\cos\theta\frac{\partial}{\partial\rho} = (a^2 - \rho^2)^{1/2}\frac{\partial}{\partial\rho} = ag(\rho)\frac{\partial}{\partial\rho},$$

where $g(\rho) = \sqrt{1 - \rho^2/a^2}$. Hence

$$\frac{1}{a^2\sin\theta}\frac{\partial}{\partial\theta}\left(\sin\theta\frac{\partial}{\partial\theta}\right) = \frac{g(\rho)}{\rho}\frac{\partial}{\partial\rho}\left(\rho g(\rho)\frac{\partial}{\partial\rho}\right).$$

In the limit $\rho/a \to 0$, we have $g(\rho) \to 1$, and the 2D Laplacian in polar coordinates emerges.

Chapter 5

5.2   We have $g_{11} = f(\theta)$, $g_{22} = 1$, $x_1 \equiv r$, $x_2 \equiv \theta$. Substitute in and you will find that most of the terms vanish, leaving

$$K = \frac{1}{2f}\left[-\frac{d^2 f}{d\theta^2} + \frac{1}{2f}\left(\frac{df}{d\theta}\right)^2\right].$$

You can try and find a function that makes this zero by actually solving the equation, but guessing may be more effective and you should quickly come across $f(\theta) = \theta^2$ as an example. In fact, such a metric is just two-dimensional polar coordinates with the letters written round the wrong way (the letters of course being arbitrary labels that can be switched at will), so you should know that that is what you need to guess.

5.5   Noting that $(a^\mu b^\nu)$ is a two-index contravariant tensor, apply the definition plus the fact that partial derivatives obey the product rule. This gives

$$
\begin{aligned}
(a^\mu b^\nu)_{;\rho} &= (a^\mu b^\nu)_{,\rho} + \Gamma^\mu_{\sigma\rho} a^\sigma b^\nu + \Gamma^\nu_{\tau\rho} a^\mu b^\tau, \\
&= \left(a^\mu_{,\rho} + \Gamma^\mu_{\sigma\rho} a^\sigma\right) b^\nu + a^\mu \left(b^\nu_{,\rho} + \Gamma^\nu_{\tau\rho} b^\tau\right), \\
&= a^\mu_{;\rho} b^\nu + a^\mu b^\nu_{;\rho}.
\end{aligned}
$$

Another way to look at this is that demanding the product rule tells us the law for differentiating rank-2 tensors.

5.7   We write three different versions of the definition of the covariant derivative of $g_{\mu\nu}$ as follows

$$
\begin{aligned}
g_{\alpha\beta,\mu} &= \Gamma^\nu_{\alpha\mu} g_{\nu\beta} + \Gamma^\nu_{\beta\mu} g_{\alpha\nu}; \\
g_{\alpha\mu,\beta} &= \Gamma^\nu_{\alpha\beta} g_{\nu\mu} + \Gamma^\nu_{\mu\beta} g_{\alpha\nu}; \\
-g_{\beta\mu,\alpha} &= -\Gamma^\nu_{\beta\alpha} g_{\nu\mu} - \Gamma^\nu_{\mu\alpha} g_{\beta\nu},
\end{aligned}
$$

where in each case it has been set to zero. Add them up and group terms taking advantage of the symmetry $g_{\mu\nu} = g_{\nu\mu}$ to obtain

$$
\begin{aligned}
g_{\alpha\beta,\mu} + g_{\alpha\mu,\beta} - g_{\beta\mu,\alpha} \\
= \left(\Gamma^\nu_{\alpha\mu} - \Gamma^\nu_{\mu\alpha}\right) g_{\nu\beta} + \left(\Gamma^\nu_{\alpha\beta} - \Gamma^\nu_{\beta\alpha}\right) g_{\nu\mu} + \left(\Gamma^\nu_{\beta\mu} + \Gamma^\nu_{\mu\beta}\right) g_{\alpha\nu}.
\end{aligned}
$$

Symmetry of the Christoffels lets us cancel the first two terms on the right-hand side, and combine the final one, giving

$$g_{\alpha\beta,\mu} + g_{\alpha\mu,\beta} - g_{\beta\mu,\alpha} = 2g_{\alpha\nu}\Gamma^{\nu}_{\beta\mu}.$$

Finally, multiply through by $g^{\alpha\gamma}/2$ and the answer emerges.

5.9 Note that the product rule is valid for an expression such as $(a^{\mu}b_{\mu})_{;\rho}$ because of Problems 5.4 and 5.6. To get the desired answer, we expand this expression in two ways.

$$(a^{\mu}b_{\mu})_{;\rho} = (a^{\mu}b_{\mu})_{,\rho} = a^{\mu}_{,\rho}b_{\mu} + a^{\mu}b_{\mu,\rho},$$

because the product is a scalar, and

$$(a^{\mu}b_{\mu})_{;\rho} = a^{\mu}_{,\rho}b_{\mu} + \Gamma^{\mu}_{\rho\nu}a^{\nu}b_{\mu} + a^{\mu}b_{\mu;\rho},$$

using the product rule and the formula for the derivative of a contravariant 4-vector. Equating the two right-hand sides, and interchanging the μ and ν dummy indices in the Christoffel term, gives

$$a^{\mu}b_{\mu;\rho} = a^{\mu}b_{\mu,\rho} - a^{\mu}\Gamma^{\nu}_{\mu\rho}b_{\nu}.$$

Since a_{μ} is an arbitrary vector, the required relation must be true for all b_{μ}.

5.10 Basically this is just an algebraic slog, but it requires some care to get all of the terms. The main trick is to get all the terms in the double covariant derivative (though sometimes sloppy calculators get lucky and omit terms that would later have cancelled anyway!). Doing the α derivative first

$$a^{\mu}_{;\alpha;\beta} = \left(a^{\mu}_{,\alpha} + \Gamma^{\mu}_{\alpha\nu}a^{\nu}\right)_{;\beta}.$$

Now the β derivative has to be done on both the indices; both terms are rank-2 tensors. So

$$\begin{aligned}
a^{\mu}_{;\alpha;\beta} = {} & a^{\mu}_{,\alpha,\beta} + \Gamma^{\mu}_{\alpha\nu,\beta}a^{\nu} + \Gamma^{\mu}_{\alpha\nu}a^{\nu}_{,\beta} \\
& + \Gamma^{\mu}_{\beta\tau}\left[a^{\tau}_{,\alpha} + \Gamma^{\tau}_{\alpha\nu}a^{\nu}\right] - \Gamma^{\gamma}_{\alpha\beta}\left[a^{\mu}_{,\gamma} + \Gamma^{\mu}_{\gamma\nu}a^{\nu}\right].
\end{aligned}$$

The first three terms are the partial derivative with respect to β, using the product rule on the second term. The first square-bracket term is the Christoffel term arising from the covariant derivative with respect to the μ index, and the second one with respect to the α index. So, you should have seven terms in all.

Then $a^{\mu}_{\ ;\beta;\alpha}$ is just the same with α and β interchanged. Line up the terms and you'll find they all cancel except the four we want. All terms in the expression are currently multiplied by a^{ν}; to get the relation we want we have to note that it holds for arbitrary a^{ν}.

5.13   To contract to obtain a rank-2 tensor, we need to sum over two indices. As there are a total of four indices, there are six different possible combinations (first with second, third, and fourth, second with third and fourth, third with fourth). Note there is no difference in which of the two summed indices is raised.

We now demonstrate that they are all zero or effectively the Ricci tensor. Four are equal to $\pm R_{\mu\nu}$ by antisymmetry, viz

$$-R^{\alpha}_{\ \mu\nu\alpha} = R_{\mu}^{\ \alpha}_{\ \nu\alpha} = -R_{\mu}^{\ \alpha}_{\ \alpha\nu} = R^{\alpha}_{\ \mu\alpha\nu} \equiv R_{\mu\nu}\,.$$

The other two are identically zero, by antisymmetry under swapping of summed indices. For example

$$R^{\alpha}_{\ \alpha\mu\nu} = -R_{\alpha}^{\ \alpha}_{\ \mu\nu} = -R^{\alpha}_{\ \alpha\mu\nu} \quad \Longrightarrow \quad R^{\alpha}_{\ \alpha\mu\nu} = 0\,.$$

Here the first step uses antisymmetry to swap the indices, generating a minus sign, while the second step raises one and lowers the other using the metric, restoring the initial form without introducing a compensating minus sign. A similar calculation shows that $R_{\mu\nu}^{\ \ \alpha}_{\ \alpha}$ is also identically zero.

5.14   The hint to specialise to a local inertial frame suggests that we use equation (5.64), and then use the fact that the resulting tensor equation is true in all frames. The short way to do the question is just to accept the equation, and differentiate to find

$$R_{\mu\nu\alpha\beta,\tau} = \frac{1}{2}\left(g_{\mu\beta,\nu,\alpha,\tau} - g_{\nu\beta,\mu,\alpha,\tau} - g_{\mu\alpha,\nu,\beta,\tau} + g_{\nu\alpha,\mu,\beta,\tau}\right);$$

$$R_{\mu\nu\tau\alpha,\beta} = \frac{1}{2}\left(g_{\mu\alpha,\nu,\tau,\beta} - g_{\nu\alpha,\mu,\tau,\beta} - g_{\mu\tau,\nu,\alpha,\beta} + g_{\nu\tau,\mu,\alpha,\beta}\right);$$

$$R_{\mu\nu\beta\tau,\alpha} = \frac{1}{2}\left(g_{\mu\tau,\nu,\beta,\alpha} - g_{\nu\tau,\mu,\beta,\alpha} - g_{\mu\beta,\nu,\tau,\alpha} + g_{\nu\beta,\mu,\tau,\alpha}\right).$$

Using the commutativity of partial derivatives, one can find that every term can be paired with an equal one of opposite sign; for example, the first term in the first equation cancels with the third term in the last equation. Hence, when the equations are added, one finds

$$R_{\mu\nu\alpha\beta,\tau} + R_{\mu\nu\tau\alpha,\beta} + R_{\mu\nu\beta\tau,\alpha} = 0.$$

One can replace the partial derivatives with covariant ones in the local inertial frame, and as it is a tensor equation, it is true in all frames. The required result can be obtained by raising the μ index.

The proof of equation (5.64) is part of Problem A3.2.

Chapter 6

6.1

a) To show that $\Lambda^\mu{}_\nu$ is a Lorentz boost in the direction of v^i, we apply it to a coordinate $x^\nu = (ct, \mathbf{x})$,

$$\begin{pmatrix} \gamma & \gamma v_j/c \\ \gamma v^i/c & \delta^i_j + (\gamma-1)\hat{v}^i\hat{v}_j \end{pmatrix} \begin{pmatrix} ct \\ x^j \end{pmatrix} = \begin{pmatrix} \gamma ct + \gamma \mathbf{v} \cdot \mathbf{x} \\ x^i + (\gamma-1)\hat{v}^i\hat{v} \cdot \mathbf{x} + \gamma v^i t/c \end{pmatrix}.$$

Noting that $x_\parallel = \hat{v} \cdot \mathbf{x}$ is the component of $\mathbf{x}$ in the direction of $\mathbf{v}$, and that $x^i_\perp = x^i - \hat{v}^i\hat{v} \cdot \mathbf{x}$ is orthogonal to $\mathbf{v}$, we can write

$$\Lambda x = \begin{pmatrix} \gamma\left(ct + x_\parallel v\right) \\ \gamma\left(x_\parallel + ct\right)\hat{v}^i \end{pmatrix} + \begin{pmatrix} 0 \\ x^i_\perp \end{pmatrix}.$$

Hence we see the form of a Lorentz boost, but only on the component along the vector $\mathbf{v}$. Taking into account the positive signs in the first term, we see the boost is in the direction $-\mathbf{v}$.

b) Apply Λ to the stationary perfect-fluid matrix, taking care to get the indices right

$$T'^\mu{}_\sigma = \Lambda^\mu{}_\nu T^\nu{}_\rho (\Lambda^T \eta)^\rho{}_\sigma.$$

We will use i, j, k, l to refer to the space components of μ, ν, ρ, σ, and write

$$\beta^i = \frac{v^i}{c}, \quad \Delta^i{}_j = \delta^i_j + (\gamma-1)\hat{v}^i\hat{v}_j$$

to save space in the following. Then we have

$$
\begin{aligned}
T'^{\mu}{}_{\sigma} &= \begin{pmatrix} \gamma & \gamma\beta_j \\ \gamma\beta^i & \Delta^i{}_j \end{pmatrix} \begin{pmatrix} -\rho c^2 & \\ & \delta^j_k p \end{pmatrix} \begin{pmatrix} \gamma & -\gamma\beta_l \\ -\gamma\beta^k & \Delta^k{}_l \end{pmatrix}, \\
&= \begin{pmatrix} -\gamma\rho c^2 & (\gamma\beta_k)\,p \\ -(\gamma\beta^i)\,\rho c^2 & \Delta^i{}_k p \end{pmatrix} \begin{pmatrix} \gamma & -\gamma\beta_l \\ -\gamma\beta^k & \Delta^k{}_l \end{pmatrix}, \\
&= \begin{pmatrix} -\gamma^2(\rho c^2 + p v^2/c^2) & -(\gamma^2\beta_l)(\rho c^2 + p) \\ -(\gamma^2\beta^i)(\rho c^2 + p) & \delta^i_l p + \gamma^2\beta^i\beta_l(\rho c^2 + p) \end{pmatrix},
\end{aligned}
$$

where we have used $\beta_k\Delta^k{}_l = \gamma\beta_l$ and $\Delta^i{}_k\Delta^k{}_l = \delta^i_l + \gamma^2\beta^i\beta_l$. Noting that $\gamma^2 v^2/c^2 = \gamma^2 - 1$, the 00 component can be written as $\gamma^2(\rho c^2 + p) - p$, and

$$
T'^{\mu}{}_{\sigma} = \begin{pmatrix} -\gamma^2(\rho c^2 + p) & -(\gamma^2\beta_l)(\rho c^2 + p) \\ -(\gamma^2\beta^i)(\rho c^2 + p) & \gamma^2\beta^i\beta_l(\rho c^2 + p) \end{pmatrix} + \begin{pmatrix} p & \\ & \delta^i_l p \end{pmatrix}.
$$

This can be rewritten in terms of 4-velocity $u^{\mu} = \gamma c(1, \beta^i)$ as

$$
T'^{\mu}{}_{\sigma} = \left(\rho + \frac{p}{c^2}\right) u^{\mu} u_{\sigma} + \delta^{\mu}_{\sigma} p,
$$

matching equation (6.26).

6.6

a) Write out the covariant conservation equation in the form $T^{\mu\nu}{}_{;\nu} = 0$ using partial derivatives and Christoffel symbols.

$$
T^{\mu\nu}{}_{;\nu} = T^{\mu\nu}{}_{,\nu} + \Gamma^{\mu}{}_{\sigma\nu} T^{\sigma\nu} + \Gamma^{\nu}{}_{\sigma\nu} T^{\mu\sigma} = 0.
$$

b) Starting from equation (5.12)

$$
\Gamma^{\mu}{}_{\nu\sigma} = \frac{1}{2} g^{\mu\rho} \left(\partial_{\sigma} g_{\rho\nu} + \partial_{\nu} g_{\rho\sigma} - \partial_{\rho} g_{\nu\sigma}\right),
$$

with the given metric components

$$
g_{00} = -\left(1 + \frac{2\Phi}{c^2}\right); \quad g_{ij} = \left(1 - \frac{2\Phi}{c^2}\right)\delta_{ij},
$$

we find the non-zero derivatives are

$$
g_{00,\rho} = -\frac{2\Phi_{,\rho}}{c^2}; \quad g_{ij,\rho} = -\frac{2\Phi_{,\rho}}{c^2}\delta_{ij}.
$$

Hence, we could write

$$g_{\mu\nu,\rho} = -2I_{\mu\nu}\frac{\Phi_{,\rho}}{c^2},$$

where $I_{00} = 1$ and $I_{ij} = \delta_{ij}$, with $I_{0j} = 0 = I_{i0}$. To first order in Φ/c^2,

$$\Gamma^\mu_{\nu\sigma} = \eta^{\mu\sigma}\left(-I_{\sigma\nu}\frac{\Phi_{,\lambda}}{c^2} - I_{\sigma\lambda}\frac{\Phi_{,\nu}}{c^2} + I_{\nu\lambda}\frac{\Phi_{,\sigma}}{c^2}\right).$$

'Slowly-changing' means we can neglect $\dot{\Phi}$ in comparison to $c|\nabla\Phi|$. This means we can drop Γ^0_{00}. Now look at Christoffel symbols with spatial components in each of the three positions:

$$\Gamma^m_{00} = \eta^{ms}I_{00}\frac{\Phi_{,s}}{c^2} = \frac{\Phi_{,m}}{c^2};$$

$$\Gamma^0_{0m} = \Gamma^0_{m0} = \eta^{00}\left(-I_{00}\frac{\Phi_{,m}}{c^2}\right) = \frac{\Phi_{,m}}{c^2};$$

$$\Gamma^m_{nr} = \eta^{ms}\left(-I_{sn}\frac{\Phi_{,r}}{c^2} - I_{sr}\frac{\Phi_{,n}}{c^2} + I_{nr}\frac{\Phi_{,s}}{c^2}\right)$$

$$= -\frac{1}{c^2}\left(\delta^m_n\Phi_{,r} + \delta^m_r\Phi_{,n} - \delta_{nr}\Phi^{,m}\right).$$

First examine the $\mu = 0$ component of the energy–momentum conservation equation.

$$0 = T^{0\nu}_{\ ,\nu} + 2\Gamma^0_{n0}T^{n0} + \left(\Gamma^0_{l0} + \Gamma^n_{ln}\right)T^{0l};$$

$$0 = T^{0\nu}_{\ ,\nu} + \frac{2}{c^2}\Phi_{,n}T^{n0} + \frac{1}{c^2}\left[\Phi_{,l} - \left(\delta^n_n\Phi_{,l} + \delta^n_l\Phi_{,n} - \delta_{nl}\Phi^{,n}\right)\right]T^{0l};$$

$$0 = T^{0\nu}_{\ ,\nu} + \frac{2}{c^2}\Phi_{,n}T^{n0} - \frac{2}{c^2}\Phi_{,l}T^{0l};$$

$$0 = T^{0\nu}_{\ ,\nu} = \frac{\partial}{\partial t}\left(\rho c^2\right) + \partial_j\left[\left(\rho c^2 + p\right)v^j\right].$$

Hence this equation is not changed.

Now examine the $\mu = i$ component of the energy–momentum conservation equation.

$$0 = T^{i\nu}_{\ ,\nu} + \Gamma^i_{00}T^{00} + \Gamma^i_{nr}T^{nr} + \left(\Gamma^0_{l0} + \Gamma^n_{ln}\right)T^{il}$$

$$= T^{i\nu}_{\ ,\nu} + \frac{1}{c^2}\left[\Phi^{,i}T^{00} - (\delta^i_n\Phi_{,r} + \delta^i_r\Phi_{,n} - \delta_{nr}\Phi^{,i})T^{nr} - 2\Phi_l T^{il}\right].$$

As we assume $\Phi/c^2 \ll 1$, $v/c \ll 1$, and $p \ll \rho c^2$, we drop all terms involving products of $\Phi_{,m}$ and $T^{\mu\nu}$, apart from T^{00},

$$T^{i\nu}{}_{,\nu} + \frac{1}{c^2}\Phi^{,i}T^{00} = 0;$$

$$\frac{\partial}{\partial t}\left(\rho v^i\right) + \partial_j\left(\rho v^i v^j + \delta^{ij}p\right) + \frac{1}{c^2}\Phi^{,i}\left(\rho c^2\right) = 0.$$

Using the 0 component of the equation, and dividing by ρ, we arrive at the required result

$$\frac{\partial}{\partial t}v^i + \left(v^j\partial_j\right)v^i + \partial^i p + \Phi^{,i} = 0.$$

Chapter 7

7.3   Start from the vacuum Einstein equation in the form

$$R^{\mu\nu} - \frac{1}{2}g^{\mu\nu}R + g^{\mu\nu}\Lambda = 0.$$

In order to reach the desired form, we need to eliminate R. This is done by contracting this equation with $g_{\mu\nu}$, from which we get

$$R - \frac{1}{2}4R + 4\Lambda = 0, \implies R = 4\Lambda.$$

The required form $R^{\mu\nu} = g^{\mu\nu}\Lambda$ follows immediately on substitution for R.

7.4   Start from the expression for the Riemann tensor.

$$R^{\mu}{}_{\nu\alpha\beta} = \Gamma^{\mu}{}_{\beta\nu,\alpha} - \Gamma^{\mu}{}_{\alpha\nu,\beta} + \Gamma^{\mu}{}_{\alpha\rho}\Gamma^{\rho}{}_{\beta\nu} - \Gamma^{\mu}{}_{\beta\rho}\Gamma^{\rho}{}_{\alpha\nu}.$$

In this weak-field metric, in Problem 6.6 we showed that

$$\Gamma^{\mu}{}_{\nu\lambda} = \eta^{\mu\sigma}\left(-I_{\sigma\nu}\frac{\Phi_{,\lambda}}{c^2} - I_{\sigma\lambda}\frac{\Phi_{,\nu}}{c^2} + I_{\nu\lambda}\frac{\Phi_{,\sigma}}{c^2}\right),$$

where $I_{00} = 1$ and $I_{ij} = \delta_{ij}$, with $I_{0j} = 0 = I_{i0}$. In the weak-field approximation, we need only the linear terms in the metric. We multiply by c^2 to reduce writing.

The derivatives of the connection are

$$c^2 \Gamma^\mu_{\beta\nu,\alpha} = -\eta^{\mu\sigma} \left(I_{\sigma\beta} \Phi_{,\nu,\alpha} + I_{\sigma\nu} \Phi_{,\beta,\alpha} - I_{\beta\nu} \Phi_{,\sigma,\alpha} \right);$$
$$c^2 \Gamma^\mu_{\alpha\nu,\beta} = -\eta^{\mu\sigma} \left(I_{\sigma\alpha} \Phi_{,\nu,\beta} + I_{\sigma\nu} \Phi_{,\alpha,\beta} - I_{\alpha\nu} \Phi_{,\sigma,\beta} \right).$$

Subtracting and lowering the first index,

$$c^2 R_{\mu\nu\alpha\beta} \simeq -I_{\mu\beta} \Phi_{,\nu,\alpha} + I_{\mu\alpha} \Phi_{,\nu,\beta} + I_{\beta\nu} \Phi_{,\mu,\alpha} - I_{\alpha\nu} \Phi_{,\mu,\beta}.$$

The required results for R_{0n0b} and R_{mnab} follow by setting $\mu\nu\alpha\beta = 0n0b$ or $mnab$, and dividing again by c^2.

The 00 component of the Einstein equation is

$$R^{00} - \frac{1}{2} g^{00} R = \frac{8\pi G}{c^4} T^{00} = \frac{8\pi G}{c^2} \rho.$$

On the left-hand side, R is already linear in the metric perturbation Φ, so we can take $g^{00} = -1$. We need the components of the Ricci tensor, so we contract the Riemann tensor on the first and third indices

$$c^2 R_{\nu\beta} = c^2 R^\alpha_{\ \nu\alpha\beta} = -I^\alpha_{\ \beta} \Phi_{,\nu,\alpha} + I^\alpha_{\ \alpha} \Phi_{,\nu,\beta} + I_{\beta\nu} \Phi^{,\alpha}_{\ ,\alpha} - I_{\alpha\nu} \Phi^{,\alpha}_{\ ,\beta}.$$

With $\nu\beta = 00$, and dropping all terms with a time derivative, as appropriate for the Newtonian approximation,

$$c^2 R_{00} = I_{00} \Phi^{,a}_{\ ,a} = \nabla^2 \Phi.$$

Hence, given that $R^{00} = R_{00}$ in this approximation, the required result that $R^{00} = \nabla^2 \Phi / c^2$ follows.

Chapter 8

8.1

a) Use the expression for the Riemann tensor,

$$R^\theta_{\ \phi\theta\phi} = \Gamma^\theta_{\phi\phi,\theta} - \Gamma^\theta_{\theta\phi,\phi} + \Gamma^\theta_{\theta\rho} \Gamma^\rho_{\phi\phi} - \Gamma^\theta_{\phi\rho} \Gamma^\rho_{\theta\phi}.$$

The only non-zero element in the sums over ρ is $\rho = \phi$ in the last term. Hence

$$R^\theta_{\ \phi\theta\phi} = (-\sin\theta \cos\theta)_{,\theta} - 0 + 0 - (-\sin\theta \cos\theta) \frac{\cos\theta}{\sin\theta} = \sin^2\theta,$$

and so

$$R^{\theta\phi}{}_{\theta\phi} = g^{\phi\phi} R^{\theta}{}_{\phi\theta\phi} = \frac{1}{r^2 \sin^2\theta} \sin^2\theta = \frac{1}{r^2}.$$

This is identical to the Riemann curvature of a two-sphere. This is to be expected, as the surfaces of constant (r,t) are two-spheres.

b) You can make things easier by noticing that $A'/A = -B'/B$, which considerably simplifies the expression to be evaluated. Then

$$B = 1 - \frac{2GM}{rc^2} \quad \Longrightarrow \quad B' = \frac{2GM}{r^2c^2} \quad \Longrightarrow \quad B'' = -\frac{4GM}{r^3c^2}.$$

So, since $AB = 1$

$$R^{tr}{}_{tr} = g^{rr} R^{t}{}_{rtr} = -\frac{1}{2} B'' = \frac{2GM}{r^3c^2}.$$

So at the Schwarzschild radius it equals $c^4/4G^2M^2$ which is finite, while it is infinite at $r = 0$. The finiteness of the curvature at the Schwarzschild radius is a strong indication that the divergence of the metric may be due to a coordinate singularity rather than a real singularity.

8.2   This requires a lot of careful algebra and we won't show all the steps. From

$$r = \tilde{r}\left(1 + \frac{GM}{2c^2\tilde{r}}\right)^2,$$

we can obtain

$$dr = d\tilde{r}\left(1 + \frac{GM}{2c^2\tilde{r}}\right)\left(1 - \frac{GM}{2c^2\tilde{r}}\right).$$

Substitute that in and begin simplifying. After lots of judicious cancelling, your aim is to reach the form

$$
\begin{aligned}
ds^2 &= -c^2\left(\frac{1 - GM/2c^2\tilde{r}}{1 + GM/2c^2\tilde{r}}\right)^2 dt^2 \\
&+ \left(1 + \frac{GM}{2c^2\tilde{r}}\right)^4 \left[d\tilde{r}^2 + \tilde{r}^2 d\theta^2 + \tilde{r}^2 \sin^2\theta\, d\phi^2\right].
\end{aligned}
$$

This is known as the 'isotropic form', because the spatial part of the metric factorises neatly (albeit at the cost of an ugly temporal part).

Now carry out an expansion for small $GM/2c^2\tilde{r}^2$ to find

$$ds^2 \simeq -c^2 \left(1 - \frac{2GM}{c^2\tilde{r}}\right) dt^2 + \left(1 + \frac{2GM}{c^2\tilde{r}}\right) \left[d\tilde{r}^2 + \tilde{r}^2 d\theta^2 + \tilde{r}^2 \sin^2\theta \, d\phi^2\right],$$

which matches the weak-field limit expression (7.10) with $\Phi = -GM/c^2\tilde{r}$.

8.3 The vacuum Einstein equation is $R_{\mu\nu} = \Lambda g_{\mu\nu}$. We can use the same combination of R_{tt} and R_{rr} as in the text to derive

$$\frac{A}{Bc^2} R_{tt} + R_{rr} = -\Lambda \left(\frac{A}{Bc^2} g_{tt} + g_{rr}\right) = 0,$$

and hence it is still true that $B'/B = -A'/A$. The $\theta\theta$ equation then becomes

$$-\frac{r}{A}\left[-\frac{A'}{A} - \frac{1}{r}(A-1)\right] = \Lambda r^2.$$

Some algebra leads to

$$\frac{d}{dr}\left(\frac{r}{A}\right) = 1 - \Lambda r^2.$$

This can be integrated to give

$$A^{-1}(r) = 1 - \frac{a}{r} - \frac{1}{3}\Lambda r^2,$$

where a is an integration constant, which can be identified with r_S. For the range of radii in the question, the metric is close to the flat metric, and geodesic motion is almost equivalent to motion in a Newtonian potential

$$\Phi = -\frac{r_S}{2r} - \frac{1}{6}\Lambda r^2,$$

and hence a radial acceleration

$$a_r = -\Phi' = -\frac{r_S}{2r^2} + \frac{1}{3}\Lambda r.$$

Thus, for $r > (3r_S/2\Lambda)^{1/3}$, the acceleration is outward. Hence in this metric gravity appears repulsive at sufficient distances.

8.5

a) The Christoffel symbols containing ϕ and ψ and their derivatives are

$$\Gamma^r_{rr} = \psi'; \quad \Gamma^t_{tr} = \Gamma^t_{rt} = \phi'; \quad \Gamma^r_{tt} = c^2\phi' e^{2(\phi-\psi)};$$
$$\Gamma^r_{\theta\theta} = -re^{-2\psi}; \quad \Gamma^r_{\phi\phi} = -r\sin^2\theta\, e^{-2\psi}.$$

b) The non-zero Ricci tensor components are

$$R^t_{\ t} = -e^{-2\psi}\left[\phi'' + (\phi')^2 - \phi'\psi' + \frac{2}{r}\phi'\right];$$
$$R^r_{\ r} = e^{-2\psi}\left[\phi'' + (\phi')^2 - \phi'\psi' - \frac{2}{r}\psi'\right];$$
$$R^\phi_{\ \phi} = R^\theta_{\ \theta} = -\frac{e^{-2\psi}}{r}\left[\phi' - \psi' - \frac{1}{r}\left(e^{2\psi} - 1\right)\right].$$

The resulting Ricci scalar is

$$R = -2e^{-2\psi}\left[\phi'' + (\phi')^2 - \phi'\psi' + \frac{2}{r}(\phi' - \psi') - \frac{1}{r^2}\left(e^{2\psi} - 1\right)\right].$$

It is now straightforward to check that the Einstein tensor is as given in the question, as $G^\mu_{\ \nu} = R^\mu_{\ \nu} - \frac{1}{2}\delta^\mu_\nu R$.

Chapter 9

9.2   First we examine

$$\frac{D_t^2 v^\mu}{d\lambda^2} = \frac{d}{d\lambda}\left[\frac{dv^\mu}{d\lambda} + \Gamma^\mu_{\ \nu\rho} t^\nu v^\rho\right] + \Gamma^\mu_{\ \nu\rho}\left[\frac{dv^\rho}{d\lambda} + \Gamma^\rho_{\ \sigma\tau} t^\sigma v^\tau\right].$$

Using

$$\frac{dt^\nu}{d\lambda} = -\Gamma^\nu_{\ \sigma\tau} t^\sigma t^\tau, \quad \frac{d}{d\lambda}\Gamma^\mu_{\ \nu\rho} = \Gamma^\mu_{\ \nu\rho,\sigma} t^\sigma,$$

we find

$$\frac{D_t^2 v^\mu}{d\lambda^2} = \frac{d^2 v^\mu}{d\lambda^2} + 2\Gamma^\mu_{\ \nu\rho} t^\nu \frac{dv^\rho}{d\lambda}$$
$$+ \Gamma^\mu_{\ \nu\rho,\sigma} t^\sigma t^\nu v^\rho + \left(\Gamma^\mu_{\ \sigma\tau}\Gamma^\tau_{\ \nu\rho} - \Gamma^\mu_{\ \rho\tau}\Gamma^\tau_{\ \nu\sigma}\right) t^\sigma t^\nu v^\rho,$$

where some index relabelling has been performed on the last two terms. We can use equation (9.18) to replace the first two terms on the right-hand side by $-\Gamma^\mu_{\nu\rho,\sigma} t^\nu t^\rho v^\sigma$. After exchanging the indices σ and ρ the derivatives of the Christoffel symbol combine to complete the Riemann tensor, and so

$$\frac{D^2_t v^\mu}{d\lambda^2} = R^\mu{}_{\nu\rho\sigma} t^\nu t^\sigma v^\rho .$$

9.4

a) To find the radii of the circular orbits, for which $dr/d\tau = 0$, we solve $dV^{GR}_{eff}(r)/dr = 0$. It is convenient to define $u = r_S/r_\circ$, in which case

$$\frac{d}{dr} V^{GR}_{eff}(r) = \frac{c^2}{2r_\circ}\left(1 - \lambda^2 u + 3\lambda^2 u^2\right) = 0.$$

The answer follows via a bit of algebra.

b) Let the period for the astronaut be T_{ast}, obtained by integrating $\ell = r_\circ^2 d\phi/d\tau$.

$$T_{ast} = \frac{2\pi r_\circ^2}{\lambda r_S c}, \quad \text{with } \lambda^{-2} = 2u - 3u^2 .$$

Time for a distant observer t is related to the time of the orbiting astronaut τ through $dt = (dt/d\tau)d\tau$. The factor $dt/d\tau$ is obtained from the velocity equation $(dx/d\tau)^2 = -c^2$, with $dr/d\tau = 0$ as the orbit is circular.

$$T_\infty = (1-u)^{-1} U T_{ast}, \quad \text{with } U^2 = c^2(1-u)(1+\lambda^2 u^2) .$$

This can be rewritten as

$$T^2_\infty = \frac{8\pi^2 r_\circ^3}{r_S c^2},$$

which is the same relation between period and orbital radius as for an orbit in Newtonian gravity.

9.6

a)

$$V^{GR}_{eff} = \frac{\Lambda \ell^2}{6} - \frac{GM}{r} + \frac{\ell^2}{2r^2} - \frac{GM\ell^2}{c^2 r^3} - \frac{1}{6}\Lambda c^2 r^2 .$$

The stable orbit disappears for large enough Λ.

b) The outward acceleration due to the cosmological constant is approximately $\Lambda c^2 r/3$ for a non-relativistic body. The radius of the Earth's orbit is approximately $r_E \simeq 1.5 \times 10^{11}$ m, and the Solar mass approximately $M_\odot \simeq 2.0 \times 10^{30}$ kg. Hence, the outward acceleration is $\left(1.1 \times 10^{-52}\right) \times \left(1.5 \times 10^{11}\right) \times \left(3 \times 10^8\right)^2 /3 = 4.9 \times 10^{-25}$ ms^{-2}. This is to be compared to the Newtonian acceleration of $GM_\odot/r_E^2 \simeq \left(6.7 \times 10^{-11}\right) \times \left(2.0 \times 10^{30}\right) / \left(1.5 \times 10^{11}\right)^2 = 5.9 \times 10^{-3}$ ms^{-2}.

For a star at $r_* = 30$ kiloparsecs outside a galaxy of mass $10^{11} M_\odot$, the acceleration due to the cosmological constant is 3.1×10^{-15} ms^{-2}. The Newtonian acceleration is 1.6×10^{-11} ms^{-2}.

Chapter 10

10.3   Consider a ray of light from S passing through the lens placed at point B with impact parameter b, and reaching O. The segments SL and LO can be taken straight. Let ψ be the angle BSL. Elementary geometry shows that $\theta + \psi = \delta\,\phi_{\mathrm{def}}$, and hence that $\theta \simeq \delta\,\phi_{\mathrm{def}} D_{\mathrm{ls}}/D_{\mathrm{s}}$, as all angles are small. Using the deflection angle formula $\delta\,\phi_{\mathrm{def}} = 2r_{\mathrm{S}}/b$, and that $b = \theta D_1$, the result follows.

10.5   Set up coordinates x', y', z', with x' and y' in the plane of the pulsar orbit, and x along the major axis. The origin is the centre of the ellipse. The semi-major axis is a_1. Then $x' = a_1 \cos E$ and $y' = a_1 \sin E$ (see Advanced Topic A2), where E is the angle to the centre of the ellipse. The orbital plane is tilted around the major axis by angle ι. Hence, in the sky-plane coordinates x, y, z, the coordinate z in the direction of the observer is $z = y' \sin \iota = a_1 \sin \iota \sin E$. Now note the relationship between E and the orbital angle ϕ (A2.19). Almost there! The final step is to rotate the ellipse in its own plane by angle w, which is defined to be the angle between periastron and the intersection of the orbital plane and the sky plane. Then $y' = a_1 \sin E \cos w + a_1 \cos E \sin w$. The change in the light travel time is $\Delta_R = z/c$, so

$$\Delta_R = \frac{a_1}{c} \sin E \cos w + a_1 \cos E \sin w = \frac{a_1 \sin \phi \cos w + a_1 \cos \phi \sin w}{1 + \varepsilon \cos \phi}.$$

Equation (10.18) is recovered with a trigonometric identity.

10.6   A moving clock measures proper time T with

$$dT^2 = dt^2 \left[(1 + 2\Phi) - (1 - 2\Phi)\frac{1}{c^2}\frac{d\mathbf{x}^2}{dt^2}\right].$$

To second order in v/c, we have

$$\frac{dT}{dt} = 1 + \Phi - \frac{1}{2}\frac{v^2}{c^2}.$$

The potential is

$$\Phi = \Phi_{1s} + \Phi_2(r),$$

where Φ_{1s} is the constant potential at the pulsar surface, and $\Phi_2 = -GM_2/rc^2$, where we neglect the pulsar radius in comparison to the distance between them. Get formula for velocity of pulsar 1 in Kepler orbit,

$$v^2 = \frac{GM_2^2}{M_1 + M_2}\left(\frac{2}{r} - \frac{1}{a}\right).$$

The result follows.

Chapter 11

11.3 The new metric is related to the old one by

$$\tilde{g}_{\mu\nu} = \frac{\partial x^\rho}{\partial \tilde{x}^\mu}\frac{\partial x^\sigma}{\partial \tilde{x}^\nu}g_{\rho\sigma},$$

so it is more direct to rewrite the transformation as

$$ct = c\tilde{t} \mp \left[\tilde{r} - r_S \ln\left|\frac{\tilde{r}}{r_S} - 1\right|\right].$$

As θ and ϕ are unchanged, we need only look at the r and t metric components.

$$\tilde{g}_{\tilde{t}\tilde{t}} = \frac{\partial t}{\partial \tilde{t}}\frac{\partial t}{\partial \tilde{t}}g_{tt}, \quad \tilde{g}_{\tilde{t}\tilde{r}} = \frac{\partial t}{\partial \tilde{t}}\frac{\partial t}{\partial \tilde{r}}g_{tt}, \quad \tilde{g}_{\tilde{r}\tilde{r}} = \frac{\partial t}{\partial \tilde{r}}\frac{\partial t}{\partial \tilde{r}}g_{tt} + \frac{\partial r}{\partial \tilde{r}}\frac{\partial r}{\partial \tilde{r}}g_{rr}.$$

The differentials are

$$\frac{\partial t}{\partial \tilde{t}} = 1, \quad \frac{\partial t}{\partial \tilde{r}} = \mp\left(1 - \frac{r_S}{r}\right)^{-1}, \quad \frac{\partial r}{\partial \tilde{r}} = 1.$$

It follows that

$$\tilde{g}_{\tilde{t}\tilde{t}} = -c^2\left(1 - \frac{r_S}{r}\right), \quad \tilde{g}_{\tilde{t}\tilde{r}} = \mp 1, \quad \tilde{g}_{\tilde{r}\tilde{r}} = 0.$$

Hence, the space–time interval is

$$ds^2 = -c^2 \left(1 - \frac{r_S}{r}\right) d\tilde{t}^2 \mp 2d\tilde{t}d\tilde{r} + \tilde{r}^2 d\Omega^2 .$$

A desirable feature is that the metric is not singular at $r = r_S$.

11.5

a) The time interval according to the crew near the black hole is the proper-time interval of a worldline with constant (r, θ, ϕ), which we denote $d\tau$. The coordinate t in the usual Schwarzschild metric is the time measured by distant observers, so

$$dt = \left(1 - \frac{r_S}{r}\right)^{-1/2} d\tau = \sqrt{2}d\tau.$$

The pulses are therefore received about 1.4 seconds apart.

b) We need to solve the radial infall equation with zero angular momentum, or

$$\mathscr{E} = \frac{1}{2}\left(\frac{dr}{d\tau}\right)^2 - c^2\frac{r_S}{2r},$$

with $\mathscr{E} = c^2 r_S/2r_0 = c^2/4$. Hence,

$$\frac{dr}{d\tau} = c\sqrt{\frac{r_S}{2r}}\sqrt{1 - \frac{r}{r_0}}.$$

The r integral can be done with the substitution $\cos\theta = \sqrt{r/r_0}$, to give

$$c\tau = \sqrt{\frac{2r_0}{r_S}}r_0 \left[\cos^{-1}\left(\sqrt{\frac{r}{r_0}}\right) + \sqrt{\frac{r}{r_0}\left(1 - \frac{r}{r_0}\right)}\right].$$

The proper time when the ship reaches $r = r_S$ starting from rest at $r_0 = 2r_S$ is therefore

$$c\tau = 4r_S \left[\cos^{-1}\left(\frac{1}{\sqrt{2}}\right) + \frac{1}{2\sqrt{2}}\right] \simeq 4.56\,r_S .$$

The pulses are emitted with proper time interval 1 second. Hence, with $r_S = 10$ light seconds, 45 pulses are emitted. The proper time when the ship reaches $r = 0$ is

$$c\tau = 2\pi r_S \simeq 6.28\, r_S\,.$$

Hence, with $r_S = 10$ light seconds, 62 pulses are emitted, although the last 17 don't make it out of the black hole.

11.6

a) Circular orbits are at radii which are solutions to $dV_{\text{eff}}^{\text{GR}}/dr = 0$, where $V_{\text{eff}}^{\text{GR}}(r)$ is given by equation (9.29). This leads to

$$\frac{r_S}{r} = \frac{1}{3}\left[1 \pm \sqrt{1 - \frac{3}{\lambda^2}}\right],$$

with $\lambda = \ell/r_S c$. This has real solutions for $\lambda \geq \sqrt{3}$. Hence, the minimum specific angular momentum is obtained from $\lambda = \sqrt{3}$ or

$$\ell = \sqrt{3}r_S c\,,$$

for which the orbital radius is $r_\circ = 3r_S$.

b) The formulae have already been worked out in Problem 9.4. Substituting $r_\circ = 3r_S$ into the expressions for T_{ast} and T_∞, we get

$$T_{\text{ast}} = \sqrt{27}\,\frac{2\pi r_S}{c}\,, \qquad T_\infty = \sqrt{54}\,\frac{2\pi r_S}{c}\,.$$

11.10 Obtain the total energy emitted by multiplying the power output per unit area by the surface area of the black hole, and using the Hawking temperature given in equation (11.12). The corresponding mass loss is given via $E = mc^2$. Then set up a differential equation for dM/dt of the form

$$\frac{dM}{dt} = -\text{const.} \times \frac{1}{M^2}\,,$$

where you will need to keep track of and numerically evaluate the constant to obtain the answer as given in the question.

A black hole of 10 Solar masses has a lifetime of 2×10^{67} years, vastly longer than even the age of the Universe. For a lifetime comparable to the Universe's age, the required initial mass is 1.7×10^{11} kg. This latter is tiny by astronomical standards, being more like the mass of a mountain than of a planet or star.

Chapter 12

12.1   The neatest way to solve the equation is to notice that

$$\left(-\frac{r}{A}\right)' = r\frac{A'}{A^2} - \frac{1}{A},$$

which combines the terms containing A into a single differential to give

$$\left(\frac{r}{A}\right)' = 1 - 3kr^2,$$

allowing immediate integration.

The Riemann tensor components are given in equation (12.2). Taking the latter and substituting the above solution we find

$$R^{\theta\phi}_{\ \theta\phi} = \frac{1}{r^2 A}(A-1) = k - \frac{C}{r^3},$$

which diverges at the origin unless $C = 0$. The other terms give an equivalent result, which for instance follows immediately via equation (12.3).

However in principle even this is not really enough, since divergence of Riemann tensor components can be due to coordinate singularities, see Problem 12.3.

12.5   Note that (x^1, x^2, x^3) look like ordinary 3D spherical polar coordinates with radial coordinate $r = \rho \sin \chi$. Hence,

$$(dx^1)^2 + (dx^2)^2 + (dx^3)^2 + (dx^4)^2 = dr^2 + r^2(d\theta^2 + \sin^2\theta \, d\phi^2) + (dx^4)^2.$$

Now, $dr = \sin\chi d\rho + \rho\cos\chi d\chi$ and $dx^4 = \cos\chi d\rho - \rho\sin\chi d\chi$. Hence,

$$(dx^4)^2 = d\rho^2 + \rho^2 d\chi^2 + \rho^2 \sin^2\chi (d\theta^2 + \sin^2\theta \, d\phi^2).$$

On surfaces of constant ρ we have

$$dl^2 = \rho^2 d\chi^2 + \rho^2 \sin^2\chi (d\theta^2 + \sin^2\theta \, d\phi^2).$$

12.6   The Friedmann equation immediately gives us

$$\frac{\dot{a}}{a} = \pm\sqrt{\frac{\Lambda}{3}}.$$

This is a linear equation whose solution is

$$a(t) \propto \exp\left(\pm\sqrt{\frac{\Lambda}{3}}t\right).$$

The plus and minus signs correspond (respectively) to expanding and contracting solutions, which are time reversals of each other due to the $t \leftrightarrow -t$ invariance of the Friedmann equation.

[Notes: de Sitter space can be represented in a number of very different coordinate systems. The most important two are as a 4D hyperbolic slice through a 5D Minkowski space–time, and as a static, spherically symmetric solution which is the special case of the Schwarzschild–de Sitter solution of Problem 8.3 with r_S set to zero. The case of negative Λ gives rise to anti-de Sitter space, which has proven of considerable interest in fundamental physics modelling, but it does not have a representation as a cosmological solution.]

Chapter 13

13.1   Following the same procedure that we used for matter or radiation, the density evolves as $\rho \propto 1/a^{3(1+w)}$ and the corresponding expansion rate is $a \propto t^{2/3(1+w)}$. These are easily checked to reduce to the matter and radiation solutions for $w = 0$ and $1/3$ respectively.

13.3   The rewriting of the Friedmann equation follows immediately from rewriting the densities in terms of their values at equality. Using conformal time $H = \dot{a}/a = ca'/a^2$, where prime is a derivative with respect to conformal time, makes the equation

$$a'^2 = \frac{1}{2}\frac{H_{eq}^2 a_{eq}^2}{c^2}\left(aa_{eq} + a_{eq}^2\right).$$

This is a separable equation, which if you wish could be brought to a minimal form via further redefinitions $\hat{a} = a/a_{eq}$ and $\eta = H_{eq}a_{eq}\tau$. The solution (it is pretty fiddly to get the integration constant in the right form) is

$$\frac{a(\tau)}{a_{eq}} = \left(2\sqrt{2}-2\right)\left(\frac{\tau}{\tau_{eq}}\right) + \left(1-2\sqrt{2}+2\right)\left(\frac{\tau}{\tau_{eq}}\right)^2,$$

where

$$\tau_{eq} = \frac{(4-2\sqrt{2})c}{a_{eq}H_{eq}}.$$

There is no good way to write this using cosmic time t. Note that the solution smoothly transitions between the separate regimes for radiation and matter found in Problem 13.2.

13.6   The age for this cosmology is given by

$$H_0 t_0 = \int_0^1 \frac{da}{a\sqrt{1 - \Omega_m + \Omega_m a^{-3}}} = \int_0^\infty \frac{dz}{(1+z)\sqrt{1 - \Omega_m + \Omega_m(1+z)^3}}.$$

Using $y = (1+z)^{-3/2}$, so that $dy/y = -(3/2)dz/(1+z)$

$$H_0 t_0 = \frac{2}{3}\frac{1}{\sqrt{1 - \Omega_m}}\int_0^1 \frac{dy}{\sqrt{y^2 + \Omega_m/(1 - \Omega_m)}},$$

from which the answer follows since $\int_0^1 (y^2 + c)^{-1/2}$ is $\sinh^{-1}(1/\sqrt{c})$ for positive c:

$$H_0 t_0 = \frac{2}{3}\frac{1}{\sqrt{1 - \Omega_m}}\sinh^{-1}\left[\sqrt{\frac{1 - \Omega_m}{\Omega_m}}\right].$$

For instance by plotting this function, we find the age equals the Hubble time when $\Omega_m \simeq 0.26$, which as it happens is quite close to the observed value.

13.8   For a matter-dominated Universe with $k = 0$ and $\Lambda = 0$, we have $a(t) \equiv (1+z)^{-1} = (t/t_0)^{2/3}$ where the scale factor is normalised to unity at present. Hence

$$r = \int_{\text{emit}}^{\text{obs}} \frac{c\,dt}{a(t)},$$

so

$$r_0 = 3ct_0\left[1 - (t/t_0)^{1/3}\right] = 3ct_0\left[1 - \frac{1}{\sqrt{1+z}}\right],$$

where the second equation assumes reception at the current time and emission at an arbitrary redshift z.

The above expression is in comoving units, so to obtain the angle subtended by an object of physical length l we must use its comoving length $l/a = (1+z)l$. Since light rays travel radially the angular size is simply

$$\tan\frac{\theta}{2} = \frac{(1+z)l}{2r_0} = \frac{l}{3ct_0}\frac{1+z}{1 - 1/\sqrt{1+z}}.$$

(Usually in cosmology the small-angle approximation would be valid and we could set $\tan\theta/2 \simeq \theta/2$.)

Then θ tends to π, i.e. occupying the whole width of the field of view, at both $z = 0$ and $z = \infty$. The first is because an object held arbitrarily close to you is obviously going to look huge. The second is because the object is of fixed *physical* size, and at early times the Universe was arbitrarily small and hence the object arbitrarily large with respect to it.

Differentiating the z-dependent part of the above formula and setting the answer to zero gives the minimum size at $z = 5/4$ (the numerator of the derivative being $2z - 3\sqrt{1+z} + 2$).

Index

Introducing General Relativity, First Edition. Mark Hindmarsh and Andrew Liddle.

© 2022 John Wiley & Sons Ltd. Published 2022 by John Wiley & Sons Ltd.

Companion website: www.wiley.com/go/hindmarsh/introducingGR